Doing Theology in the New Normal

Doing Theology in the New Normal

Edited by Jione Havea

scm press

Published in 2021 by SCM Press
Editorial office
3rd Floor, Invicta House,
108–114 Golden Lane,
London EC1Y 0TG, UK
www.scmpress.co.uk

SCM Press is an imprint of Hymns Ancient & Modern Ltd
(a registered charity)

H
Y   Ancient
M
N   &Modern
S

Hymns Ancient & Modern® is a registered trademark of
Hymns Ancient & Modern Ltd
13A Hellesdon Park Road, Norwich,
Norfolk NR6 5DR, UK

British Library Cataloguing in Publication data

A catalogue record for this book is available
from the British Library

978-0-334-06064-2

Typeset by Regent Typesetting
Printed and bound by
CPI Group (UK) Ltd

Contents

Part 3 In Decencies

Part 4 In Protest

Acknowledgement

Work on this book was supported by
the Council for World Mission

Foreword

This publication is another step along the journey of 'discernment and radical engagement' (DARE, a programme of the Council for World Mission) with God's mission, this time in the context of the Covid pandemic.

Covid changed the world. We were required to observe physical distancing and to work, if not to also 'live', in virtual space. The revelations of Covid, a 'pandemic of inequality', were stark – including the endemic societal issues of racism, poverty and gender-based violence – all of which are closely tied to the contest to maintain the 'empire's economy'.

While the pandemic forced us to look at the brokenness of our society, it also made us witnesses to the celebrations of flora and fauna, of earth breathing, of trees clapping, of air becoming lighter and cleaner, of waters teeming with life – made possible by our absence. Covid is therefore an opportunity for us to also focus on life-flourishing perspectives, and to grapple with disruptions without losing faith or weakening our grip on the alternatives to which our aspirations and discernment lead. To this end, we are not looking to return to the old normal; instead, we want to follow where God leads so that together we may engage the Covid disruptive ways of living with the confidence and courage that faith in God enables.

It has been said that Covid is another manifestation of capitalism. I agree. The debate on when or how to reopen the economy seems to be over human life versus the economy. Such a dichotomy is unhealthy. What we need is not a debate on life versus economy but the affirmation of the *economy of life* and the aspiration for *life-flourishing communities*.

Covid disrupted life and in its wake infected over 98 million persons and left over 2.1 million dead (by the middle of January 2021). Those who were infected by the coronavirus have brought us face to face with our mortality, liminality and vulnerability. We all share the pain of not being able to offer the usual ritual of homecoming to our loved ones; and, while we offer our sympathies to those who have gone through this frightening experience, we live with the nagging fear that any of us could be the next to go through this trauma.

And yet we must live with the pain of irresponsible personal behaviour and poor examples from those who occupy the offices of leadership. The callous disregard and disrespect by many of our leaders towards individuals and families who have suffered and are suffering the effects of this coronavirus is both unbelievable and unacceptable.

This publication explores the question of normalcy, on the one hand, and the 'new normal', or what I choose to call the unfolding future, on the other. Two positions are emerging, and they are not mutually exclusive:

1 stubborn resistance to allow the disruptive forces of this pandemic to define the basis of life in the unfolding future nor to determine our destiny;
2 strong impulse to see this pandemic as one among many, some of which are preventable but are not given the attention they deserve perhaps because of social location.

In consideration of these positions the critical concerns remain the same – the economy, the ecology, systemic racism and its consequent violent oppression against black and brown people, xenophobia, homophobia, white privilege and patriarchy.

The contributors to this publication engage with the fierceness of the Covid pandemic and reflect on what it has exposed. The key questions that this work puts before us are: What can we as God-talking people offer in a time of pandemic, social disintegration and broken systems? What resources can we summon to promote a future that is peace-affirming, justice-focused and life-flourishing?

The Council for World Mission (CWM) believes that *rising to life* is the strategic path for now. Over the past ten years our focus was on resisting empire, highlighting its evil and challenging its injustice. Today we declare that we will not allow the forces of death and destruction to command our lives and our movements.

We will not become servants of Covid. On the contrary, by God's grace we will become creators of a new pathway to life, a new way of being and behaving, a new pattern of living based on justice and peace. We will pursue this imagination to sharpen expectation on what a different world could be, to deepen our appreciation for the resources that are available, and to build resilience for participation in God's promised New Heaven and New Earth.

CWM is grateful to the contributors who have taken the time to prepare and to offer perspectives to inform these conversations. The conversations in this work are insightful and instructive, stimulating and hopeful.

I invite readers to allow our views to contend in this daring God-talk, confronting our vulnerabilities, acknowledging the ambiguities and complexities of life, embracing the intersections of our lived experiences and wrestling with God as we seek to make sense of a God who is sovereign and vulnerable, omnipotent and incarnate, on the throne and on the cross. In the midst of fear about resurgence and lockdowns, inadequacy of health-care facilities and lack of political will, we as people of faith must be ready to wrestle with God, to stay on the side of life flourishing.

Collin I. Cowan
General Secretary, Council for World Mission

Preface

Covid has infected what we do, how we do them and with whom. It has pushed us to work, worship, play and think differently, and even to construct and use new terms and concepts. In this publication, we adopt the following terms (the cases are intentional):

- 'Covid-19' refers to the novel coronavirus (SARS-COV-2) and the disease it causes, first reported to WHO on 31 December 2019.
- 'Covid' refers to the pandemic that WHO declared on 11 March 2020, caused by Covid-19.
- 'Covid' is also a qualifier for Covid-19 effects (e.g., covid denial, covid politics, covid times, etc.).

At the height of the first covid wave in Europe and the Americas, and in the midst of busy lives and frustrating lockdowns, the contributors to this monograph presented the first draft of their essays in a series of webinars (30 October to 1 November 2020) at which they received feedback from one another and from a wider audience. The essays were then revised and submitted for review, revision, and the editing and publication processes. Not all of the presentations are included in this publication, but the included essays and verses attempt to 'do theology in the new normal'. Insofar as the 'new normal' is still waving, this collection is one step in that drive.

The webinar series was held under the title 'Daring God-talk: what is normal?' and it was dubbed 'eDARE 2020' (see https:// edare.cwmission.org) – 'e' because the gathering was held in virtual space, and DARE (Discernment and Radical Engagement) is a programme of the Council for World Mission (CWM) that includes an annual Global Forum (that has met face-to-face since 2017). The DARE Global Forum could not be held in 2020 because of Covid, hence eDARE 2020. This publication brings the daring theological and cultural responses of the contributors, and their conversation partners and communities, to Covid, and this publication was 'done' thanks to their time and wisdom.

A publication like this happens, also, thanks to the support of many others. Collin I. Cowan, general secretary of CWM, has given generously of his energy and affection to DARE and now, in addition, to eDARE. *Praise Jah*.

Many thanks also to David Shervington and the SCM Press editorial board and team, for welcoming this project and guiding the processes of publication.

Lastly, but not finally, to the eDARE team – Michael Jagessar, Maria Fe (Peachy) Labayo and Sainimili Kata – and to the energetic and energizing CWM team of Julie Sim, Faris Ariffin, Stephen Chia, Simeon Cheok and Agnes Chen: *salamat* and *vinaka saka*.

About the Contributors

Juliana Claassens is professor in Old Testament and head of the Gender Unit at the Faculty of Theology, Stellenbosch University. She is the author of *Writing/Reading to Survive: Biblical and Contemporary Trauma Narratives in Conversation* (Sheffield Phoenix, 2020); *Claiming Her Dignity: Female Resistance in the Old Testament* (Liturgical, 2016); *Mourner, Mother, Midwife: Reimagining God's Liberating Presence* (Westminster John Knox, 2012) and *The God who Provides: Biblical Images of Divine Nourishment* (Abingdon, 2004).

Delana C. Corazza is a social scientist who researches around evangelicals and politics with the Tricontinental Institute for Social Research. Author and co-author of academic publications, magazines, book chapters, newspaper articles, and websites on the outskirts, human and social rights, and religion. Worked on a project by the Pontifical Catholic University against forced evictions, coordinated academic research on evictions in the city of São Paulo and Santo André, and coordinated a project with women in social vulnerability.

Wanda Deifelt is an ordained Lutheran pastor and theologian who is professor at Luther College (Decorah IA, since 2004), and visiting faculty at Emmanuel College (Toronto, Canada) and Gurukul (Chennai, India). She served as professor of Feminist Theology at Escola Superior de Teologia (now Faculdades EST, São Leopoldo) for 14 years. At EST she also served as vice-president, dean of graduate studies, and national coordinator of International Network of Advanced Theological Education. Her publications are in the areas of constructive and contextual theologies, embodiment, ecumenical and interreligious dialogue.

Dorothea Erbele-Küster is senior scholar in Biblical Literature, Gender and Diversity at the Johannes Gutenberg University in Mainz, Germany. Her research focuses on biblical ethics and anthropology, feminist and intercultural hermeneutics. Among her publications is *Body, Gender and Purity in Leviticus 12 and 15* (T&T Clark, 2017). As co-editor of a new series, *Theologische Interventionen*, she recently published on the intersections of ethics and aesthetics (Verführung zum Guten, 2018).

Beverley Gail Haddad is senior research associate at the School of Religion, Philosophy and Classics, University of KwaZulu-Natal, South Africa. She has worked in the field of the church and development for the past 30 years, both as a researcher and as an ordained priest in the Anglican Church of Southern Africa. She has published extensively, with her most recent work focusing on the theological challenges posed by the increasing rate of HIV infection among young African women and the emerging Covid pandemic.

Jione Havea is co-parent for an active and talkative six-year-old, migrant to the Wurundjeri land of the Kulin Nations, native pastor (Methodist Church in Tonga) and research fellow with Trinity Methodist Theological College (Aotearoa New Zealand) and with the Public and Contextual Theology research centre (Charles Sturt University, Australia). Jione authored *Losing Ground: Reading Ruth in the Pacific* (2021) and co-edited with Monica J. Melanchthon *Bible Blindspots: Dispersion and Otherings* (2021).

Wei Huang is assistant professor at the Department of History of Shanghai University, China. Her interests include ancient Israelite history and its ANE background, cross-textual interpretation of the Hebrew Bible and Chinese Classics. Her recent publication in English is 'Hebel and Kong: A Cross-textual Reading between Qoheleth and the Heart Sūtra', in Athalya

Brenner-Idan, Archie Chi-chung Lee and Gale A. Yee (eds), *The Five Scrolls: Texts@Contexts*, pp. 134–44 (Bloomsbury, 2018).

Volker Küster is professor of Comparative Religion and Missiology, Johannes Gutenberg-Universität Mainz, Germany. Küster explores the interconfessional, intercultural and interreligious dimensions of Christian faith with methods of hermeneutics, aesthetics, communication, postcolonial and globalization theories. His research evolves along two lines: dialogue, conflict and reconciliation and visual art and religion. His recent publications include *A Protestant Theology of Passion: Korean Minjung Theology revisited* (Brill, 2010); *God / Terror. Ethics and Aesthetics in Contexts of Conflict and Reconciliation* (Equinox, 2021).

Tat-siong Benny Liew is class of 1956 professor in New Testament Studies at the College of the Holy Cross, USA. He is the author of *Politics of Parousia* (Brill, 1999) and *What Is Asian American Biblical Hermeneutics?* (UCLA Asian American Studies Center, 2008). His edited works include *Postcolonial Interventions* (Sheffield Phoenix, 2009), *They Were All Together in One Place* (with Randall Bailey and Fernando Segovia; Brill, 2009), *Reading Ideologies* (Sheffield Phoenix, 2011), *Present and Future of Biblical Studies* (Brill, 2018), and *Colonialism and the Bible: Contemporary Reflections from the Global South* (with Fernando Segovia; Lexington Books, 2018).

Sung Uk Lim is associate professor of New Testament at Yonsei University, South Korea. His research focuses on the construction of Jewish and Christian identities from a multi-ethnic and gender-inclusive perspective. He is particularly interested in the intersections of race, ethnicity, gender, sexuality and postcolonialism in order to enhance a multifaceted understanding of religions in the ancient Mediterranean world. His book *Otherness and Identity in the Gospel of John* (Palgrave, 2021) examines the narrative construction of identity and otherness in the Gospel of John.

Tinyiko Maluleke is professor of Theology and senior research fellow at the Centre for the Advancement of Scholarship, University of Pretoria, South Africa. He is a member of the South African Academy of Science and a regular contributor of op-ed pieces in the South African media. His recent publications include: 'Racism *En Route*. An African Perspective', *Ecumenical Review* 72.1 (2020): 19–36; 'Black and African Theologies in Search of Comprehensive Environmental Justice', *Journal of Theology for Southern Africa* 167 (2020): 5–19.

Michael Mawson is from Aotearoa New Zealand and lives in Sydney, Australia. He is senior lecturer in Theology at the School of Theology, and research fellow in the Public and Contextual Theology Research Centre, both at Charles Sturt University; and research fellow for the Theology for Southern Africa Initiative at the University of the Free State, South Africa. He is the author of *Christ Existing as Community: Bonhoeffer's Ecclesiology* (Oxford University Press, 2018).

Kuzipa Nalwamba is a member of staff at the World Council of Churches, teaching Ecumenical Social Ethics at Bossey Ecumenical Institute and is the Programme Executive for Ecumenical Theological Education (ETE). She is a retired ordained minister of the United Church of Zambia. Her dissertation at the University of Pretoria was in Dogmatics and Christian Ethics, with a focus on eco-theology. She remains a keen student and researcher in the field, with particular interest in the interdisciplinary intersection of Christian, scientific and cultural perspectives.

Keun-joo Christine Pae is associate professor of Religion/Ethics and Women's and Gender Studies at Denison University, Granville, Ohio. She is also chair of the Religion Department at the college. Trained as a social ethicist, her research and teaching include womanist/feminist spiritual activism, faith-based popular activism, transnational feminist ethics, religious ethics of peace and war, US overseas military bases and military prostitution, and Asian/Asian American feminist theology and ethics.

Anna Kasafi Perkins is a senior programme officer with the Quality Assurance Unit in The University of the West Indies (UWI) and adjunct faculty at St Michael's Theological College, Jamaica. She teaches and researches in ethics, justice, popular culture, sexuality, theology, scripture, and quality assurance. She serves on the UWI Ethics Committee and the UWI Covid-19 Task Force. Her most recent publication is *Ethics Amidst Covid-19: A Brief Ethics Handbook for Caribbean Policy-makers and Leaders* (2020), co-authored with Professor R. Clive Landis.

James W. Perkinson (PhD University of Chicago) has lived for 35 years as a settler on Three Fires land in inner city Detroit, currently teaching Social Ethics at Ecumenical Theological Seminary. He is the author of five books including *Political Spirituality in an Age of Eco-Apocalypse: Communication and Struggle Across Species, Cultures, and Religions; Shamanism, Racism, and Hip-Hop Culture: Essays on White Supremacy and Black Subversion*. He is an artist on the spoken-word poetry scene and long-time activist.

Anthony G. Reddie is the director of the Oxford Centre for Religion and Culture, Regent's Park College, University of Oxford, and a professor extraordinarius with the University of South Africa. He is a prolific author, having written a number of books, journal articles and book chapters. He is also the editor of *Black Theology: An International Journal*. He is a leading international researcher having been awarded an 'A' rating in the National Research Foundation in South Africa, and is a recipient of the 2020 Lanfranc award by the Archbishop of Canterbury for his exceptional contribution to Black Theology.

Hadje Cresencio Sadje is an associate member in the Centre for Palestine Studies, University of London (UK). He holds a Master of Arts in Ecumenical Studies (Sociology of Religion) at the University of Bonn (Germany) and has worked with various professional and faith-based organizations including

Christian Peacemaker Team, Caritas Brussels, Peace Builders Community Philippines and Pananaw Pinoy. He is currently pursuing a PhD in Religious Studies at Hamburg University (Germany) under his advisor Professor Giovanni Maltese.

Karen Georgia Thompson is an ordained minister in the United Church of Christ (USA). She serves as the associate general minister for Wider Church Ministries, and the co-executive for Global Ministries, the joint mission agency of the United Church of Christ and the Christian Church (Disciples of Christ). She is one of the elected officers of the denomination. She is a writer and poet who has published widely. Her poetry reflects her passion for justice and the fight for human rights. Her first book of poetry *Drums in Our Veins* is scheduled for publication in 2021.

Angelica Tostes is a feminist theologian, MA in Religion Studies at the Methodist University of São Paulo. She researches with Tricontinental Institute for Social Research, teaches in a post-graduation program of Youth, Citizenship and Religion at the Catholic Faculty of Santa Catarina. Author and editor of books and articles about feminist theology, religion and politics, interfaith dialogue, and multiple religious belonging. Interfaith activist and a member of the Global Interfaith Network for People of All Sexes, Sexual Orientation, Gender Identity, and Expression.

Upolu Lumā Vaai is principal of the Pacific Theological College (Suva, Fiji), where he also teaches subjects in theology and ethics. His research areas include the doctrine of Trinity, theologies of relationality, and Pacific indigenous relational philosophies. His publications include *Relational Hermeneutics: Decolonising the Mindset and the Pacific Itulagi* (co-edit with Aisake Casimira; University of the South Pacific and the Pacific Theological College, 2017) and *Relational Self: Decolonising Personhood in the Pacific* (co-edit with Unaisi Nabobo-Baba, 2017).

Gerald O. West is professor emeritus in the School of Religion, Philosophy and Classics in the University of KwaZulu-Natal, South Africa. He has worked extensively with the Ujamaa Centre for Community Development and Research for the past 30 years, a project in which socially engaged biblical scholars and ordinary African readers of the Bible from poor, working-class and marginalized communities collaborate for social transformation. His most recent book is *The Stolen Bible: From Tool of Imperialism to African Icon* (Brill, 2016).

Sithembiso S. Zwane is a lecturer in Bible and Social Change in the School of Religion, Philosophy and Classics, within the College of Humanities, University of KwaZulu-Natal. He is a member of the Evangelical Lutheran Church of Southern Africa (ELCSA) in the South Eastern Diocese (SED) Ondini Circuit, and he serves as the director of the Ujamaa Centre for Biblical and Theological Community Development and Research. His research interest is the intersectionality between theology, development and the Bible.

I

New but Old: Go and Do Otherwise

JIONE HAVEA

First, a prayer: May the ancestors receive the parents and grand-parents, sisters and brothers, orphans and widows, lovers and strangers, homeless and neighbours, who passed on, drowned, in the waves of Covid-19. They no longer breathe, but each one of them was named – may their names be said and remembered in the hearts, lives and actions of survivors and grievers. And may we who can still breathe say and do something about the pandemics at hand. In other words, may we who can still breathe do more than simply say 'amen'.

Next, some questions: What race were the wounded (on the side of the road) and frontline worker (at the inn) in the so-called Parable of the Good Samaritan (Luke 10.30–35)? Could they too have been Samaritans? Did they have companions or helpers? How old were they? Would they have been presented and read differently if they were of a different race, different gender, different class, different colour, different age? What would Jesus say?

On the one hand, my queries are inappropriate. The text is a parable which was told for a particular purpose – to answer the question, 'Who is my neighbour?' (Luke 10.29) – and it is (as parables tend to be) sparing of details. This was not a recounting of an actual event, so that i[1] or other readers could decide which details are factual and which are fake.

On the other hand, simply because this is a parable, my queries tease the text to life. They pry the parable from the interrogations of the young lawyer, and put the young Jesus on the spot: why didn't Jesus make the young lawyer, and many readers since, see and understand race, class, age, colour and companionship in and around the wounded 'half dead' traveller? Why didn't Jesus make the Good Samaritan return and fulfil his commitment to the innkeeper, who was stuck with a patient rather than a patron? Didn't the innkeeper too show mercy to the robbed and wounded traveller (cf. Luke 10.37)?

My queries, which may also be raised on behalf of the robbers (who are, thus designated, discriminated against by default), refuse to let the Good Samaritan, the young lawyer and/or Jesus, control how this parable is read. They have had their say, but as a parable this text says more than what they want it to say. If what those characters said are understood as versions of the 'old normal', my queries symbolically invite attending to concerns that arise with the 'new normal'.[2]

Old Normal

At the outset, Covid has been somewhat epoch-making – it instigated the setting of the 'old normal' and the dawn of the 'new normal'. The global community decided early in 2020 that, across the board, we need new ways of doing things – at home, in community and across public, domestic and national borders – so the primary response to Covid was to (pur)chase the essentials of life (which turned out to include toilet paper), and according to those – define the new normal that the pandemic has ushered in. The new continued to be essentialist like the old, and the ghost of Kohelet sighed over the global community: 'Vanity of vanities. All is vanity' (Eccles. 1.1).

As with the Millennium or Year 2000 bug (Y2K), Covid hiked fear around the world. But whereas the Y2K scare was related to computer programming expected to crash (but did not happen as proclaimed) when the calendar moved from year 1999 to year 2000, Covid was caused by a biological bug (SARS-Cov-2, Covid-19) that spread in human populations across the world from year 2019 to year 2020. Both 'bugs' incubated fear, but Covid-19 made real people (compared to real computers in the case of Y2K) sick and killed many of them. At the dawn of year 2000, Y2K proved to be superfluous;

at the dawn of year 2020, the Covid-19 virus began to mutate into several strands and to go strong, sickening and killing real people. And the rampage of Covid will continue for several years into the future.

The infectiousness of the disease – reaching over 860,000 reported new cases worldwide in one day, 7 January 2021 (*Statista* 2021) – hastened the turn to the new normal. In haste, driven by fear, the global community turned to embrace the new normal without first assessing the old normal. Many people were pushed to the new normal with the expectation that at some point, whenever Covid eases up, the world will return to the old normal. Behind their expectation is the assumption that the old ways and old practices were 'normal' and thus acceptable. However, so much in the ways and practices prior to Covid were unhealthy and unhelpful, sickening and deadly.

Covid is one pandemic among many, and it has not (at the time of writing) reached the devastation and pandemic proportion of, for example, the HIV/AIDS pandemic. This is not to say that one pandemic is worse than the other, but in order to see Covid in the frames of the old normal. By the end of 2020, seeing that the number of Covid deaths was disproportionately much higher among poor black and ethnic minorities (even within white societies), Covid began to appear very much like an endemic (disease found among particular people). Of course, the virus does not discriminate on the basis of race, colour, gender, class or sexual orientation. But providers of protection and services do discriminate, and minorities do not have much of a chance with those who discriminate. At the beginning of 2021, with the rolling out of vaccination campaigns, whole nations of black and ethnic minority people face covid-discrimination – they are not counted among the essential or vulnerable people to receive the vaccine first. Sadly, they may not even get in the queue before the end of 2021 (see Wasuka 2021). In these regards, Covid shares the same endemic temperature as HIV/AIDS (see Chapter 9 by Beverley Haddad, and Chapter 10 by Volker Küster).

There are many ways in which the old normal was discriminating, unhealthy and oppressive. We should not want to return to those kinds of situation (see Chapter 21 by Anthony Reddie). But then, did we (in the new normal) really move away from the old normal? Is the new normal not the old normal with a mask, or in a different skin? Could Covid be an opportunity to also look back in order to see what might still be useful from the old normal? In terms of theology, the hope expressed in the last question can take place in several ways.

First, it can involve interrogating problematic theologies of the old normal (see Chapter 11 by Hadje Sadje) and re-engaging theologies meant to assist recovery from pandemic-like crises (see Chapter 7 by Gerald West). Second, it can also involve rereading and problematizing the normality of some scriptural texts, and old readings of those texts, favoured in the old normal (see Chapter 2 by Sung Uk Lim, and Chapter 23 by Juliana Claassens). Third, it can also involve reinvigorating methodologies belittled in the old normal, such as indecent (see Chapter 16 by Christine Pae, and Chapter 27 by Wanda Deifelt) and liberation (see Chapter 14 by Sithembiso Zwane) criticisms. Fourth, it can also involve affirming the teachings and voices that were marginalized in the old normal (read: modernity), maybe because of their sublime non-scientific (see Chapter 6 by Wei Huang) or spiritual (see Chapter 17 by Michael Mawson) overtones. Fifth, it can also involve learning from communities on whose shoulders the old normal have stomped (see Chapter 4 by Angelica Tostes and Delana Corazza, and Chapter 18 by Upolu Vaai). Sixth, it can also involve encouraging emotions such as grief (see Chapter 24 by Tat-siong Benny Liew) and rage (see Chapter 25 by Dorothea Erbele-Küster) that are usually judged to be unacceptable in so-called civil societies. Seventh, it can also involve accepting that in the worldwide web of life human beings are insignificant (see Chapter 20 by James Perkinson), vulnerable (see Chapter 13 by Kuzipa Nalwamba) and destined for the graveyard (see Chapter 28 by Tinyiko Maluleke). It can also involve other approaches, but i suggested the seven above in order to locate the voices in this collection within the frames of the old normal.

While this collection is intentional about suggesting ways of *doing theology in the new normal*, the contributors also engage with and interrogate the old normal. Put another way, Covid is somewhat epoch-making – but epochs inter- and over-flow. As with mutations and rites of passage, for all organisms and across all communes, there is something old and something new in every ritual, in every epoch, in every pandemic, in every movement, and in every theological imagination.

(dis)Integration

Emmanuel Garibay's *Healing and Hope* (see Figure 1.1) shows a human figure in a state of disintegration or of coming together, of integration, depending on how one looks at it. In the sky is a hand among the clouds which suggests the presence of the divine, drawing upon Michelangelo's *Creation of Adam* fresco (c. 1508–12) on the ceiling of the Sistine Chapel. The hand signifies that in spite of human intelligence, the divine has a hand in the fate of humanity. Garibay allows for the divine

Figure 1.1: Emmanuel Garibay, Healing and Hope
(oil on canvas, 2016, used by permission of the artist).

hand to heal and give hope to the disintegrating body, on the one hand, and/or to judge and even to disintegrate the human composition.

On the left side is a dandelion with its dried flower spreading seeds on the barren landscape. The choice of the dandelion is deliberate as it is regarded by humans as a useless and undesirable weed. The dandelion protests against humans' arbitrary disregard for nature in terms of what agrees with their tastes, as opposed to allowing nature to take its course.

In the context of the massive environmental degradation caused mainly by human activities, this work suggests that healing and restoration of the land necessarily require repentance and disintegration of human constructions. This could be in terms of a radical change of our way of life from consumerism to prudent use of the earth's resources, from greed to compassion, from conflicts to peace and justice.

In the context of Covid, Garibay invites viewers to see the multiplexity of the divine presence and to feel the (un)hopefulness of the human condition. All of those, without losing sight of the (dis)integration of the barrened planet. Garibay's invitation is incitement for doing theology in the new normal.

At this juncture, another prayer: May the senses of smell and taste return to those who survive Covid-19 and, in another regard, may the senses of smell and taste overtake the keepers of the old normal so that they value the dandelions before them.

New Normal

For Jesus, the parable of the traveller who became a neighbour in Luke 10.30–35 was an opportunity to teach, to activate: 'Go and do likewise.' The young lawyer had to decide for himself which of the characters that passed was 'a neighbour to the man who fell into the hands of the robbers' (Luke 10.36, NRSV) and then he was to 'go and do likewise'. If he then went and did what he had learned from Jesus, a rabbi and an activist, the young man would have 'infected' his community to be good neighbours, good Samaritans.

The young lawyer was not precise in his response – the neighbour would be 'the one who showed him mercy' (Luke 10.37) – unlike many readers who assume that it is clear, and that there was only one character, who showed mercy to the robbed-and-wounded man. On the other hand, as the above musings invite, the privileged 'rich' Samaritan traveller was not the only one who showed mercy. The innkeeper also showed mercy to the robbed-and-wounded man, who may also have been among the 'rich' in his community.[3]

In the context of Covid, i suggest that it is not essential to determine who showed mercy or who was neighbourly but, in line with my reflection on the old normal, to name and hear those who have been wounded and robbed. In this context, the challenge is to let the wounded and robbed become one's neighbour. In other words, Luke 10.30–35 is also a parable of the wounded and robbed. Doing theology in the new normal thus involves shifting positionalities from being (as theologians) privileged and rich travellers or saviours (who know the truth and the way) to becoming worthy of being neighbours to the wounded and robbed 'Samaritans'. But the call is still the same, 'Go and do.' Go and activate. Incite. Infect.

I present the contributions to this collection below in the new normal frames of shifting positionalities and shifting personalities. The contributors, as a collective, invite readers to find ways in which the theologies that we do are *in touch* and *in relation* with the wounded and robbed, as well as have the courage to be *(in)decent* and *protest*.

In Touch

The robbed man, because of his placement and condition, touched the Samaritan traveller. He moved the Samaritan to touch and pour oil and wine on his wounds, and then put him on his animal and brought him to the inn (Luke 10.34). The robbed and wounded man engaged the pity of the Samaritan, and he as a consequence became a neighbour.

The five essays in this first section engage with the demand to be, and the costs of being, *in touch* with the robbed and wounded in the context of Covid. The essays flow from offering alternative readings of biblical texts to inviting critical awareness of Covid and indigenous bodies, and to reconsidering 'church theology' and its place in the economy.

Sung Uk Lim wrestles with the meaning of healing in the era of 'untact' (zero or no contact), in which safety is assured through the lack of contact with others. To prevent the spread of the coronavirus, many people have made it a habit to make no bodily contact with others. Touching neighbours, not to mention strangers, is socially banned. In this context, Lim rereads episodes of Jesus' healing acts in which he touched (e.g., leper, deaf and mute, and blind in Mark 1.40–45; 7.31–37; 8.22–26), or was touched by (e.g., a bleeding woman in Mark 5.25–34), wounded bodies.

Anna Kasafi Perkins questions the meaning of the body – 'this mortal coil' – as it has been laid bare(r) by the Covid pandemic. Drawing on cases arising from the pandemic in the Caribbean, Perkins questions the fundamental value given to bodies in space and place – dead bodies, gendered bodies, loose bodies, working bodies, poor bodies, essential bodies. How relevant and meaningful is the notion of 'mortal coil' – often mistaken as biblical, but it is Shakespearean rather – in such an exploration?

Angelica Tostes and Delana Corazza engage with the spirituality and mysticism in the Landless Workers' Movement – *Brigadas de Trabalho de Base* (grassroots brigades) – in Brazil. Facing an oppressive and neo-fascist government, this movement advocates protests and solidarity actions. Spirituality as a path to resistance is found beyond religions and their institutions; the spirituality needed to incite resistance is found in the streets, in marches, in the screams for justice and in the hands that help. This spirituality is found in the Quechua concept of *sumak kawsay, buen vivir* (good living or well living). *Buen vivir* is a community-centric idea of harmony with humans and nature.

Wei Huang juxtaposes Traditional Chinese Medicine (TCM) – based on Heaven–Human Harmony, a philosophical idea shared by Confucianism, Daoism and Buddhism – with the Hebrew Bible teaching that a good relationship with YHWH is essential for facing sufferings and crises. The theology of impurity in the Hebrew Bible works as a system of social boundaries. Huang's cross-cultural interpretations aim to diagnose and heal misunderstandings, and to establish communications in our covid infected world.

Gerald O. West looks at how Covid contributed to our understanding of the contours of the *Kairos Document* (1985) called 'church theology'. West tracks the characteristics of 'church theology' from their explication in 1985 through to Covid's 2020 enunciation of what is now the predominant form of Christian theology in South Africa. The summons of the Covid pandemic to forms of 'prophetic theology' is clear, but before we can conceptualize this task, we must interrogate the normative theologies of our time.

In the essays in this section, it is important to be *in touch*. But with whom and with what? With and for the wounded and robbed: Yes. With and for the barrened planet: Yes, in words and with deeds.

In Relation

The priest and the Levite (Luke 10.31–32), one could argue, were practising 'social distancing'. They might thereby receive the admiration if not approval of some of the irresponsible leaders in the Covid context. Could their passing by on the other side be a showing of mercy to the robbed and wounded man? Could they have gone around in order to make room for the Samaritan, who was better trained and equipped to help? Why do we, especially Christians, uncritically judge against the presumably Jewish priest and Levite? These questions are not considered in the parable, but they invite reconsideration of our readings, relations and biases.

In the context of social distancing, the five essays in this section invite interrogating relations – of Covid with the HIV pandemic, of Covid with cultural, social and theological teachings, and of Covid with the human condition of vulnerability and preferential option for solidarity. These relations do not explain the decision by the priest and Levite to pass on the other side, but they invite considering Covid in relation (rather than in isolation) to other pandemics and ideologies.

Beverley Gail Haddad intersects the ways in which the HIV and Covid pandemics in the South African context are configured in the public realm. The HIV pandemic remains private with the onus on the individual to choose to be tested and whether or not to disclose their status. This has led to a pandemic of stigma and discrimination. On the other hand, Covid is public in every way with the South African state declaring a national disaster. Haddad presents three theological imperatives (Christological, Ecclesiological and Missiological) emerging from the HIV and Covid pandemics, all of which require liberatory and prophetic responses.

Volker Küster asks why people do not always look, by default, to the Bible and to religious texts for answers to the problems of life. On the other hand, faced with Covid, people find a French philosopher of the absurd (Albert Camus) and texts from popular culture (e.g., by Susan Sontag, who became famous for her paradoxical polemics against interpretations of illness and the HIV/AIDS pandemic) more helpful. In this context, Küster's question is straightforward: Do we dare to interpret Covid with or without God?

Hadje Sadje assesses the eschatological emphasis in rapture theology popular among Pentecostal – 'a thief in the night' (1 Thess. 5.2) – movements. Sadje suggests that, to get to the heart of the oneness Pentecostal eschatology, decolonizing Christian eschatology is needed. In decolonizing Pentecostal eschatology, one becomes more attentive, sensitive and responsive to the global socio-economic and political crises of Covid.

Kuzipa Nalwamba orients towards faith-action. Covid has exposed the fault-lines and hypocrisies that run through local and global communities and asks: Had the virus been confined to the global South, would it still make global headlines? Against that backdrop, Nalwamba reflects on what it means to confess faith in a vulnerable God and to be bodily vulnerable.

The vulnerabilities that Covid exposes incite/invite the 'body' (physical and ecclesial) to lament, protest and disrupt as witness to a vulnerable God.

Sithembiso S. Zwane responds to the growing social and economic inequalities that Covid has exposed, calling for solidarity assurance for the most vulnerable. First, Zwane provides a candid reflection on the socio-economic *Reality* of Covid-19 in the South African context. Second, he argues for the 'preferential option for the poor' because *Faith* communities have not been spurred by Covid. Third, he invites a pragmatic *Action* response – such as that provided by the Ujamaa Centre for Biblical and Theological Community Development and Research, using Contextual Bible Study (CBS) and the distribution of food parcels to mitigate against the Covid pandemic.

Whereas Covid forced people to weather their illnesses and even to die in isolation, the essays in this section invite us to explore the complexes and mutations of relationality. In health and in sickness, the preferred option is not to pass on the other side but to engage and be in solidarity.

In Decencies

Could the indecencies in the parable of the wounded and robbed man be also (read as) decent, and could the decencies be indecent also? Could there be something decent in the priest, Levite or innkeeper? And could there be something indecent in the Samaritan traveller? Beyond the parable, what else are (in) decent with the young lawyer, Jesus, readers and their readings? Such are the kinds of questions that body, and queer theologies invite.

The essays in this section extend the line of interrogation favoured by body and queer theologies, which are liberational through and through, to Covid. In the context of Covid, the five essays in this section problematize the divide between decency and indecency in theological explorations.

K. Christine Pae explores how we may audaciously practise solidarity and creatively imagine peace and justice in a social distancing world. Pae calls for transnational solidarity – which is favoured by body and liberation theologies. But how would transnational solidarity look if we counter national isolationism or protectionism in the context of Covid? Because world politics is filled with warlike rhetoric, the church must practise a new ecclesial gathering. Arguably, the Christian vision of the Kin-dom of God is in tension with sovereign nations of borders. Accordingly, an alternative understanding of empathy may contribute to international solidarity.

Michael Mawson reflects on Julian of Norwich's daring, unruly God-talk that we desperately need in the context of Covid. Mawson first indicates some ways in which Julian's situation resembles the covid situation, then turns to Judith Butler's insights on performative language and excitable speech to frame his analysis of (1) Julian's intimate, sensual descriptions of Christ's bodily suffering, and (2) her fluid, destabilizing portrayals of Christ as mother. In both cases, Julian disrupts and displaces accepted (and acceptable) ways of talking about God and how God in Christ relates to us.

Upolu Lumā Vaai presents *lagimālie* (well-being) as concept for resisting the systems and categories that shaped pre-pandemic Eurocentric theologies and development narratives. Such thinking confined us to view life through separated strands to achieve one answer, one truth, or one destination. Against the 'onefication' of life and God, Vaai invites us to strive for wholistic *lagimālie* in the context of Covid. *Lagimālie* depends on how we navigate the interconnectedness of life that constitutes 'we are', in contrast to the 'we have' of market-driven neoliberal capitalism. This requires shifting from the 'one God' to the 'many [in] we' and reframing the theological questions from 'why did God punish us' to 'what can we do together'. This 'we are' approach problematizes the rationalistic direction of the theology of God and also the pietistic heavenization that dominates Pacific Christianity where everything, including pandemics, is transferred to God's juridical abode.

James W. Perkinson reflects on how Covid is not so much a killer as a revelator – revealing not some monstrous plague but unmasking humanity's toxic eco-destruction. The coronavirus is a 'coronation' – a globe-wide initiation into a new demand of evolutionary sovereignty. Perkinson, a white male settler at the curve of water called Detroit, calls for a more indigenously responsible exploration that unveils the ongoing decimations that white supremacy and settler colonialism continue to visit on populations indigenous and of colour.

Anthony G. Reddie explores the nature of protest against the prevailing clamour for a return to the old normal. At the outbreak of Covid, it was often asserted that the deaths of people from across strata of society represents the virus as 'a great leveller'. However, the disproportionate deaths of black people and those at the bottom of the socio-economic ladder show Covid as a 'great revealer'. Covid has revealed the iniquitous nature of global, neoliberal capitalism that has rendered the victims as disposable, collateral against a system that has been rendered as normal. Covid provided a painful moment of pause; an opportunity to stand back and reflect on the kind of world of which we want to be a part. The necessity of protest, grounded in the substantive theo-ethical commitments of justice found in liberation theologies, is needed if we are to challenge the knee-jerk desire to return of the 'same old, same old'. The disproportionate deaths of poor black people have shown the brokenness of the existing world order. We cannot and should not go back to the old normal.

(in)Decently, this section interflows bodily and human-centred interests with co-creaturely planetary concerns. In those flows, the essays interject at the hymen-like border between decency and indecency, calling for alternative modes of doing theology, in the spirits of solidarity, collaboration and resistance.

In Protest

In a way, the parable of the wounded and robbed man was a 'protest' against the young lawyer. Jesus just had a moment of pride, bragging about how blessed his disciples were to see things through him (Luke 10.21–24), then the young man stood to 'test' him (Luke 10.25) – it would have been an indecent moment for Jesus. In this connection, Jesus told the parable in order to put the young man in his place.

In a way also, the five essays in this section protest in order to put a few things into place. Collectively, the authors protest against the responses to, and the expectations for after, Covid.

L. Juliana Claassens addresses how Covid feels like 'an infinite present' that goes on and on, and on. The sun rises; the sun sets, only to rise again tomorrow (Eccles. 1.5). Zoom meetings. Home-schooling. Teach online. Walk around the yard. Stress baking. Watch Netflix. Repeat. We are stuck in an infinite present with no vision of tomorrow and frustrated by our inability to plan the future. But perhaps even grimmer, we have been hurled into a world of Jeremiah where we are no longer allowed to celebrate weddings. Or to enter a house of mourning where we may engage in rituals of lament and comfort of the bereaved (Jer. 16.5–9). And yet, one finds in both Ecclesiastes and the book of Jeremiah, amid a time of uncertainty and duress, traces of a theological response that refuses to accept the new normal. In both biblical books, one finds valuable resources to help us consider how to live amid the infinite present wrought by Covid.

Tat-siong Benny Liew looks back over how difficult the year 2020 was for many around the world. Things were made more difficult in the USA under the incompetent and imperialistic Trumpian regime. Trump's lassitude towards the covid crisis laid bare that he valued the economy over humanity, while his response to Black Lives Matter protests demonstrated that he would not hesitate to use not only violence or even the military but also religion to safeguard and (re)enforce his business-first policies. In the covid time of many deaths and much dread, with many people feeling what Du Bois calls 'not hopelessness but unhopeful', Liew argues that continual grief and mourning are much more than psychologically cathartic; grief and mourning can function as productive forces to insist that things can be different.

Dorothea Erbele-Küster affirms rage and lament as healthy responses to Covid, drawing upon literary responses in the Hebrew Bible and the works by contemporary poets and songwriters. She focuses on the role of the emotion of rage and the poetical mode of lament, juxtaposing recent and ancient psalms, songs and poems, and unfolds how the rage for justice are expressions of prayer and hope. Psalms, songs and poems unfold how Covid serves as a burning lens of our constant crisis. Erbele-Küster shows the interconnectedness of different poetical modes responding to crisis and highlights that lament and rage function as call to assert and to amend at the same time.

Wanda Deifelt reflects on the early view that Covid was the great equalizer. If nothing else, Covid has laid bare the inequalities that plague us also in non-pandemic times: not all workers can work remotely, not all have the right and access to unemployment benefits, not all those infected have access to health care, and not everybody has a home to self-quarantine. A common stratagem in explaining the increasing numbers in contamination and death due to Covid-19 has been to blame the victims. Emphasizing a person's conditions – such as poor health, chronic illness, obesity, homelessness, age, etc. – places an onus on the individual without acknowledging the systemic inequalities that lead to these conditions in the first place. Personal vulnerability is labelled as the culprit for massive numbers of casualties instead of naming the social and political abandonment that has relegated victims to their historically ascribed place.

Tinyiko Maluleke too resists returning to 'business as usual' post-Covid. It is bad enough to wallow in the nostalgia of a pre-Covid world, but to try and return to that world would be treasonous. Worse still would be to model the future in a pre-Covid image. The coronavirus has exposed the frailties of our immune system, the rot in our living arrangements, the deadly shortcomings in our relationship with the environment, and the poverty of our relationship with God. Covid has unmasked the world of its innocence, even as it has forced us to wear new and fragile masks. Maluleke imagines a new world, beyond the graveyard (reflecting on Mark 5) and the prison (through the work of Ben Okri).

This section asks, in the end: How might we construct a post-Covid global society and church? At some point, the

Covid pandemic will ease up and humans will learn to live with the virus. Covid-19 will mutate but, like other viruses, it will not go away. In defeat. We must therefore be resilient. And protest.

In Verses

Interspersed among the essays are verses, poetic benedictions, incited by the calls of the authors of this collection. The verses broke-in while Karen Georgia Thompson and i listened to oral presentations by the authors. Pondering their words. Breathing their troubles. Troubled by their despairs. Yet willing to climb their hills (see Gorman 2021).

With verses this work seeks, as a collective, to do theology in the new normal, as well as to do theology that inverses the new normal. On that note, a dare: Go and do otherwise.

Notes

1 I use the lowercase with the first person when i am the subject, in the same way that i use the lowercase with 'we', 'you', 'she', 'he', 'it', 'they' and 'other'. There is no justification for privileging (by capitalizing) the first person individual (a leaning of English grammar) who *is* in relation to other subjects.

2 Covid has forced me to here think in terms of the 'new normal', but my preference would be to reflect on 'other normal(s)' (which i leave for another occasion).

3 Could the robbed and wounded traveller, who would have had something to be robbed, be from the same class as the Samaritan and/or the innkeeper? The parable does not answer my question, which arises from the context of Covid where the privileged and rich are among the first to receive free care and healing.

References

Amanda Gorman, 2021, 'The Hill We Climb: The Amanda Gorman poem that stole the inauguration show', *The Guardian*, 20 January (accessed 21.1.21: www.theguardian.com/us-news/2021/jan/20/amanda-gorman-poem-biden-inauguration-transcript).

Statista, 2021, 'Number of new cases of coronavirus (Covid-19) worldwide, 23 January 2020–10 January 2021, by day', 10 January (accessed 20.2.21: www.statista.com/statistics/1103046/new-corona virus-covid19-cases-number-worldwide-by-day/).

Evan Wasuka, 2021, 'Pacific Islands may not finish rolling out Covid vaccines until 2025', *ABC News*, 21 January (accessed 21.1.21: www.abc.net.au/news/2021-01-21/pacific-islands-covid-rolling-out-vac cines-until-2025/13073054).

PART I

In Touch

2

The Touch of Jesus in a Time of *Untact*

SUNG UK LIM

Ironically, this chapter reimagines the *touch of Jesus* in this dire situation calling for *untact* – a neologism for zero contact – in an attempt to put the brakes on the spread of the Covid-19 virus. The Synoptic Gospels in particular depict Jesus as frequently touching and being intermittently touched by those suffering from critical illness for the purpose of healing.[1] In his time, Jesus' mere touch was believed to have the power to heal the sick and even the dead. His touch can thus be dubbed a *healing touch*.

The Covid pandemic is challenging and changing the world in ways that take public health as top priority. To prevent the spread of the virus, the vast majority of global citizens have made it a habit to maintain social distancing, which is designed to make lesser or no contact with others in daily life as well as in religious life. In this context, even a touch or contact with neighbours, let alone strangers, is socially and culturally banned. What is worse, all types of gathering, religious or otherwise, which may result in contacts between people, are under control for hygienic reasons. This urgent situation prompts us, Christians, to beg a knotty question: What is the meaning of divine touch in a touchless society?[2]

To respond to this question, I investigate a recurring motif of touch in biblical texts. A closer look at the episodes of Jesus' healing, especially in Mark's Gospel, shows that he sometimes strenuously touches the sick (a leper, a deaf man and a blind man; Mark 1.40–45; 7.31–37; 8.22–26) and the dead (Mark

5.21–24a, 35–43) and is, at another time, stealthily touched by a bleeding woman (Mark 5.24b–34). In the Markan perspective, the touch of Jesus can be construed as a symbol for divine salvation, or more precisely, divine healing.

This essay wrestles with what healing by divine touch means in what could be termed the society of *untact*, in which safety is believed to be guaranteed only through the sheer lack of contact with others. My contention is that Jesus' touch can be read as the signal for his boundary crossings over the social norms of purity, which extend from antiquity to the present. Furthermore, Jesus' symbolic action of touching encourages us to reimagine the meaning of a touchless touch, a new phenomenon generated in the time of *untact*, as divine intervention in human suffering in the post-Covid era.

Divergent Reconfigurations of Touch in Philosophical and Religious Imagination

Prior to analysing the healing stories of the Gospel of Mark in detail, this section undertakes a succinct survey of philosophical and religious construals of the sense of touch in general terms. To begin with, it is important to notice that the sense of touch plays a pivotal role in communicating emotions between human beings at the interpersonal level, and furthermore, between divine and human beings at the religious level. Candy Gunther Brown writes crystal clearly:

> The sense of touch is foundational to human, indeed to all animal, experiences and influences religious life. Touch is at once a physiologically based perception, grounded in receptor neurons concentrated in the skin, and evokes affective and metaphorical meanings – as in a 'touching' experience, getting 'in touch', or 'reach out and touch someone' to express relational connection and empathy. (2009, p. 770)

Brown goes further by emphasizing that touch signifies mutual affection and connection, giving rise to the feeling of empathy:

'An ability to touch or to be touched by others implies both separation and communication; touch is reciprocal – as one touches, one is touched' (p. 770). In this vein, the sense of touch is deemed constitutive of emotional transmission at the interpersonal level.

In spite of the significance of touch in terms of human relationships, Western philosophy has a long-standing tendency to denigrate sense per se in the process of seeking the truth with the conviction that sense is unceasingly liable to change and is by no means as truthful as it seems. The philosophy of Plato is a great example for this.[3] Plato casts aspersions on senses on the grounds that they are, in nature, fallacious (Brown 2009, pp. 770–1). Plato makes clear that one cannot afford to apprehend truth (ἀλήθεια) nor being (οὐσία) by means of perception (αἴσθησις), such as seeing (ὁρᾶν), hearing (ἀκούειν), smelling (ὀσφραίνεσθαι), feeling cold (ψύχεσθαι) and feeling hot (θερμαίνεσθαι): 'By which, we say, we are quite unable to apprehend truth, as we are unable to apprehend being, either' (ὧι γε, φαμέν, οὐ μέτεστιν ἀληθείας ἅψασθαι: οὐδὲ γὰρ οὐσίας) (*Theaetetus* 186d-e).[4] For Plato, sense itself is far from attaining the truth.

As far as this study is concerned, a more interesting thing to note is that touch gets less credit than the other senses. Plato hints at his hierarchical understanding of senses by praising the pleasure related to 'pure' (καθαρός) senses, say, sight, smell and hearing (*Philebus* 50e–53c). Pascal Massie contends that 'At *Philebus* 51a–52a Plato classifies the five senses into two groups; touch and taste are deemed "impure" since they are associated with needs. On the other hand, the "noble" senses (smell, hearing, and sight) deal with what remains distant and thereby are more akin to contemplation' (Massie 2013, p. 98, n.11). Likewise, Aristotle gives precedence to sight and hearing by stating that 'Of these [touch, taste, smell, hearing, vision], sight (ὄψις), on the one hand, is for the necessities and in itself; hearing (ἀκοή), on the other hand, is for the mind and by accident' (*De Sensu et Sensibilibus* 437a). This implies that Aristotle looks down on such senses as touch (ἁφή) and taste (γεῦσις), both of which are common to all animals, when compared to the other senses, that is, smell (ὄσφρησις), hearing (ἀκοή), and vision (ὄψις) (*De Sensu et Sensibilibus* 436b). Lucidly, Aristotle gives priority to sight over touch: 'sight is superior to touch' (*De Generatione et Corruptione*, 329b13). Brown (p. 771) is adamant that Aristotle in *De Anima* goes even so far as to set up the hierarchies of senses in such a way that places the sense of vision at the highest level and that of touch at the lowest level (cf. Massie 2013, pp. 74–101). Danijela Kambaksovic and Charles T. Wolfe give a witty remark: 'If sight is privileged in the idealist philosophical tradition, as the contemplation at a distance of the objects of perception, touch, the contact sense, the dirty sense, is all the way at the other extreme' (2016, p. 108).[5] In the Western tradition, probably with the exception of modern empiricism, touch has been disparaged as the most erroneous, and therefore, basest out of all five senses – sight, hearing, smell, taste and touch.[6]

On the contrary, touch, in the realm of religious thinking, is differently construed as a modus operandi to heal those suffering from illness.[7] As noted above, touch serves to communicate to others, conveying their emotions. Especially when people experience agony in the course of life, touch can function as a crucial medium for expressing empathetic feelings for the others (Brown 2009, p. 772). Furthermore, touch is also believed to be a means to alleviate suffering in the process of healing. This is deeply rooted in the belief that divine touch is able to entail pain relief, and most importantly, ultimate healing (p. 777).

Taken together, it is an irony to state that touch, which is considered as the most defective sense in philosophical thinking, plays a vital role in revealing God's willingness to participate in human sufferings with a view to generating the power of healing. With this dual aspect of touch in the philosophical and religious arenas in mind, I will scrutinize the episodes related to a reciprocal touch between Jesus and the sick in the Gospel of Mark.

Jesus' Healing Touch through Boundary Crossings in Mark's Gospel

In this section, I delve into the episodes that deal with both the event of Jesus' touch (Mark 1.40–45; 5.21–24a, 35–43; 7.31–37; 8.22–26) and that of Jesus being touched (Mark 5.24b–34) in Mark's Gospel.[8] When seen in light of Jewish purity laws, I claim that, by way of his touch (either active or passive), Jesus creates a more inclusive community by crossing over the boundaries between health and illness, between purity and impurity, between the privileged and the marginal in ancient society.[9] The implication here is that Jesus breaks the boundaries through close contact with those sick or dead, inviting a liminal space for those hygienically marginal to be integrated into a new Christian community. Let us look more closely into each episode one by one.

The first episode concerns Jesus' healing of a leper (λεπρὸς) begging him to cleanse (καθαρίσαι) his leprosy (Mark 1.40–45). In response, Jesus reaches out his hand and touches (ἥψατο) him, giving an order that he be made clean (v. 41). It comes as a surprise that Jesus' touch demonstrates the power to heal the sick.[10] It is worth noting that leprosy (λέπρα) in Jesus' time was deemed so medically infectious that a leper was forced to live alone outside the Jewish community (Lev. 13.46). Also remember that first-century Judaism held leprosy to be unclean or impure at the sociocultural and religious levels (Josephus, *A.J.* 3.261). In this light, Jesus runs the risk of contracting infectious disease and contravening Jewish purity laws by means of his warm-hearted gesture of touching a leper (Dube 2018, p. 2; Dube 2019, p. 487). In this way, Jesus himself infringes Jewish purity codes, which are intended to keep infectious diseases away, through his touch on the leprous body. It is through his touch that Jesus is rendered all the more vulnerable to contagion at the medical, sociocultural and religious levels. The final result is that the leprous man can become reincorporated into the Jewish community by Jesus' imparting his purity to him (Boring 2006, pp. 71–2).

The second episode deals with Jesus' raising of the daughter of Jairus, a leader of the synagogue (Mark 5.21–24a, 35–43). Aware that his daughter is on the verge of death, Jairus asks Jesus to come and 'lay his hands' (ἐπιθῆς τὰς χεῖρας) in order that 'she may be saved and live' (σωθῇ καὶ ζήσῃ) (v. 23). From this it follows that Jairus already knows the healing effect of Jesus' touch before his encounter with him. It is interesting to observe that Jesus does grasp (κρατήσας) the already dead body of Jairus' daughter to raise her from death (v. 41).[11] Remember that touching a corpse was considered to contravene Jewish purity code in antiquity for religious and unquestionably hygienic reasons (cf. Lev. 21.1–4; Josephus, *A.J.* 3.261). Thus, Jesus succeeds in breaking down social, cultural and religious barriers.

The third episode involves Jesus' restoration of a deaf man (Mark 7.31–37). It is shocking to see Jesus putting his fingers into his ears, spitting and touching (ἥψατο) his tongue (v. 33). Jesus adamantly crosses over the physical demarcation between himself and the deaf man by dint of his unrestrained touch. It is yet to be noticed that 'Jesus travelled extensively in Gentile territory, thus crossing boundaries he ought not to cross and exposing himself to pollution on every side' (Neyrey 1986, p. 108). No doubt, for the sake of his open ministry, Jesus does not care much about potential contamination.

The fourth episode recounts Jesus' cure of a blind man at Bethsaida (Mark 8.22–26). Jesus is called on to touch (ἄψηται) the blind man for the sake of recovering his sight (v. 22). As in the second episode, the healing power of Jesus by his touch must have been well-known to people at Bethsaida. What is interesting is that Jesus puts his saliva to the eyes of the blind man, 'laying his hands' (ἐπιθεὶς τὰς χεῖρας) on him (v. 23). Seen from the Jewish perspective, it would be a defiling action to insert one's own emissions into someone else's body in accordance with ritual purity (Neyrey 1986, p. 108).[12]

The final episode describes the healing of a haemorrhaging woman by her touch on Jesus (Mark 5.24b–34). In marked distinction from the other episodes, Jesus is stealthily touched by the woman, not vice versa. It would be astonishing to the reader, both ancient and contemporary, that even touching (ἥψατο) clothes has the power (δύναμις) to heal an illness (v. 30). This is, as in the other passage, affirmed by the request of the sick in Gennesaret: 'They [the sick] begged him to let them touch (ἅψωνται) even the edge of his [Jesus'] cloak, and whoever touched (ἥψαντο) it was healed (ἐσῴζοντο)' (Mark 6.56). In sharp contrast to Jesus' touch in the previous episodes, this time it is the haemorrhaging woman who contravenes bodily boundaries by dauntlessly touching Jesus even though she is more likely to be deemed unclean due to the flow of blood and, for this reason, is expected to go through social isolation (Lev. 15.19; cf. Josephus, *A.J.* 3.261) (Malina and Rohrbaugh 2003, p. 167). With this observation, it can also be said that Jesus, willingly or unwillingly, allows those in desperate need to come into contact with him.

All things considered, it is remarkable that Jesus heals the sick by his touch alone, and what is more, has the audacity to violate Jewish purity code in his healing ministry. The end result is that he breaks down social and cultural barriers between the healthy and the sick, between the pure and the impure, between the privileged and the marginal in first-century Jewish society. It is worth noticing that the body of the disabled, the sick and the dead is not merely a physical body but also a social body in the symbolic dimension.[13] Jerome H. Neyrey poignantly states that Jesus' revolutionary ministry towards those sick, unclean and marginal people 'replicates the lowering of purity boundaries and speaks again to the inclusive membership of Mark's community' (Neyrey 1986, p. 121). In other words, Jesus can be seen as a boundary-breaker in the Markan healing episodes by restoring individuals and a community. Warding off fear of contagious diseases, Jesus demonstrates his engaging compassion for those suffering from illness by breaking through and crossing over social and cultural boundaries by means of his intrepid touch. In other words, Jesus leverages his dauntless touch to create a more inclusive community in his time and beyond (Eck and Aarde 1993, p. 34). The transforming effect of his daring touch is to render social structures all the more malleable so that the so-called others can be reintegrated into the dominant Jewish society.

Concluding Remarks

Many, though not all, Christians worry about the future of the church in the aftermath of the Covid pandemic. They deem it nearly impossible to return to what used to be normal prior to the unprecedented outbreak of the coronavirus disease. In more general terms, Žižek makes clear:

> The catch is that, even if life does eventually return to some semblance of normality, it will not be the same normal as the one we experienced before the outbreak. Things we were used to as part of our daily life will no longer be taken for granted, we will have to learn to live a much more fragile life with constant threats. We will have to change our entire stance to life, to our existence as living beings among other forms of life. (Žižek 2020, p. 78)

It would not be an exaggeration to state that the world in general and the church in particular before and after the pandemic will be by no means the same. No matter how pessimistic we may be in the present circumstances, the coronavirus pandemic regrettably gives us a costly chance to reflect upon the quintessence of the church embodied by Jesus and his ministry on earth.

In this vein, we have thus far re-examined the meaning of touch in the philosophical and religious arenas in general and biblical texts in particular. It is suggested that comprehension of touch varies in light of philosophical and religious thinking. There remains a hidden irony that touch, a sense held to be the most basic by most philosophers, can be taken to mean divine participation in human sufferings with a view to healing those marginalized by medical systems, both ancient and contemporary, especially in the religious imagination. Frederick J. Gaiser inspiringly writes that 'healing in light of Christ is not to be untouched by pain and suffering, but to participate in Christ's own "great love," giving ourselves for others and sharing their suffering in response to Christ who bore the suffering and disease of all' (2010, p. 14).

It is through his touch of healing that Jesus breaks down social and cultural barriers between the pure and the impure, between men and women, between the wealthy and the poor, between the privileged and the marginalized, to create a more inclusive society. Remember that 'a healing touch ... bridges the isolation that many are experiencing as a result of the Covid-19 virus' (Conley 2020, p. 14).

It is incumbent on us all to mimic Jesus' indomitable gesture of getting in touch with those suffering even in the post-coronavirus era, crossing over social and cultural boundaries today (see also Hidalgo 2020; Melanchthon and Varkey 2020; Nam 2020). In Gaiser's words, Jesus' touch as a means of building relationships hints at a 'deeper touch' profoundly grounded in sharing of emotion and mutual respect in such a way as to foster a sense of dignity and mutual support (Gaiser 2010, p. 8).

All in all, following in the footsteps of Jesus and mimicking his daring *contact*, or relation-making, with the untouchable, any attempt to cross social and cultural boundaries in the time of *untact* would render a society all the more *intact* through our symbolic actions to engage the world in need of social connection, solidarity and love.

Notes

1 The Synoptic Gospels variously witness to Jesus as an agent of God who is sometimes willing to touch the untouchable, for example, the leper (Matt. 8.1–4; Mark 1.40–45; Luke 5.12–16) and at other times is touched by the impure, for instance, the bleeding woman (Matt. 9.20–22; Mark 5.24b–34; Luke 8.42b–48) for the sake of healing. These examples invite the reader to ponder the significance of touch in the time of *untact*.

2 It is natural that a thinker should reconsider the meaning of *untact* rather than contact in the setting of Covid. At the beginning of his sensational book, Slavoj Žižek raises a taxing question based on John 20.17, which is thoroughly germane to the current pandemic situation at hand: 'Today, however, in the midst of the coronavirus epidemic, we are all bombarded precisely by calls not to touch others but to isolate ourselves, to maintain a proper corporeal distance. What does this mean for the injunction "touch me not?"' (pp. 1–2). Drawing on the

significance of social distancing as a preventive measure to decelerate the spread of the novel coronavirus, Žižek shrewdly turns his attention to the Johannine resurrection scene (John 20.11–18) in which Jesus appears to Mary Magdalene with the injunction 'Do not touch me' (μή μου ἅπτου) (John 20.17). Furthermore, Žižek incisively interprets Jesus' injunction as a gesture of 'the bond of love and solidarity between people', in the contemporary world struggling to contain the pandemic (p. 77). He goes on to state that 'not to shake hands and isolate when needed IS today's form of solidarity' (p. 77, emphasis original). Even though it is of great import to consider the repercussions of Jesus' command not to touch him, this project to the contrary delves into the gist of Jesus' healing ministry grounded on touch in the so-called time of *untact*.

3 Because of space limitation, I limit the scope of this study to ancient Greek philosophy.

4 All translations are my own, unless otherwise stated. Note that Plato also deals with the issues related to the senses of touch, taste, smell, hearing and seeing in the *Timaeus* (61c3–68d7).

5 On the general survey of the sense of touch in ancient and modern philosophy, see Synnott 1991; Classen 1993; Clements 1997; Rorty and Nussbaum 2003; Magee 2000; Mattens 2009.

6 Even so, it is noteworthy to remember that Aristotle recognizes the sense of touch (ἀφή) as common to all animals (*De Anima* 413b5). Without contest, touch is one of the commonest and most indispensable senses to animals, including humanity.

7 We also find the hierarchy of senses in religion, somewhat distinguishable from that in philosophy. For instance, hearing, as a way of heeding the word of God, in the field of religion is held to be more important than any other senses. Yet, the key point is that touch was deemed to be an important way of communication between divine and human beings in early Christianity. See Barna 2007.

8 On the religious effect of touch in the Greco-Roman context, see Hughes 2018.

9 On the concept of purity and pollution in the anthropological field, see Douglas 1966. Mary Douglas is the first to point to the concept of purity as that which is in place in marked contrast to the concept of pollution as that which is out of place on the symbolic dimension governing the social and cultural classification and structure of a society. Douglas contributes significantly to a new understanding of the distinction between purity and pollution on the basis of boundary demarcation and maintenance. Jesus, seen from this perspective, would be seen as the one who bravely crosses boundaries at the social, cultural and religious levels. On the reception of Douglas' theory of purity in biblical studies, see Neusner 1973; Schwartz, Meshel, Stacker and Wright 2008; Malina 1981; deSilva 2000; Neyrey 1986.

10 Pieter J. Lallemann (1998) convincingly argues that the concept of healing by touch is characteristic of Jewish-Christian literature, having no relationship with Greco-Roman literature written before the rise of Christianity. To illustrate, Lallemann adds that the motif of healing through bodily contact is originally found in Jewish texts (1 Kings 17.21; 2 Kings 4.34; 5.11; 13.21). Needless to say, as can be seen in the present study, Christian texts bear witness to a plethora of healing episodes by Jesus and his disciples.

11 The Gospel of Mark uses two different Greek words, ἅπτεσθαι and κρατεῖν, both of which mean to 'touch' without any different nuances.

12 For an in-depth survey of the relationship between sense and its defection in the Canonical Gospels through the lens of critical disability studies, see Lawrence 2013. The greatest benefit from disability studies would be that one can think about Jesus' touch in a reciprocal sense. This approach places emphasis on the agency of the disabled in interactions with Jesus. For example, Lawrence notes: 'Transgressive reappropriations from blind perspectives, including the recovery of touch and the reordering of social relationships it involves (also sound, kinaesthesia, and speech), refigure characters not as passive recipients of healing, but rather powerful agents who, in different ways, can, at times, educate Jesus and other sighted individuals they encounter)' (p. 55).

13 Cf. Lawrence, pp. 36–7. 'The "body politic" is widely known from Graeco-Roman literature in which the physical body imaged the populace and social problems were frequently characterized as diseases. The symbolic meanings of these disease metaphors often aligned with dominant and domineering positions in conceiving of the marginal or lower classes as threatening the health of the whole polis.' See also Martin 2009.

References

Gábor Barna, 2007, 'Senses and Religion: Introductory Thoughts', *Traditiones* 36.1: 9–16.

M. Eugene Boring, 2006, *Mark: A Commentary*, Louisville: Westminster John Knox.

Candy Gunther Brown, 2009, 'Touch and American Religions', *Religion Compass* 3/4: 770–83.

Constance Classen, 1993, *Worlds of Sense: Exploring the Senses in History and across Cultures*, London: Routledge.

Ashley Clements, 2016, 'The Senses in Philosophy and Science: Five Conceptions from Heraclitus to Plato', in Jerry Toner (ed.), *A Cultural History of the Senses in Antiquity*, pp. 115–38, London: Bloomsbury Academic.

Brian Conley, 2020, 'Ministry during Covid-19: How Do We, Know-ing that We Are Susceptible to the Covid-19 Virus, Imitate Jesus in Offering a Healing Touch?' *The Priest* (May): 10–14.

David A. deSilva, 2000, *Honor, Patronage, Kinship & Purity: Unlock-ing New Testament Culture*, Downers Grove: IVP Academic.

Mary Douglas, 1966, *Purity and Danger: An Analysis of Concepts of Pollution and Taboo*, London: Routledge and Kegan Paul.

Zorodzai Dube, 2018, 'The Talmud, the Hippocratic Corpus and Mark's Healing Jesus on Infectious Diseases', *HTS Theological Studies* 74.1: 1–4.

——, 2019, 'Ritual Healing Theory and Mark's Healing Jesus: Impli-cations for Healing Rituals within African Pentecostal Churches', *Neotestamentica* 53.3: 479–89.

Evan Eck and Agvan Aarde, 1993, 'Sickness and Healing in Mark: A Social Scientific Interpretation', *Neotestamentica* 27.1: 27–54.

Frederick J. Gaiser, 2010, 'In Touch with Jesus: Healing in Mark 5.21–43', *Word & World* 30.1: 5–15.

Jacqueline M. Hidalgo, 2020, 'Scripturalizing the Pandemic', *Journal of Biblical Literature* 139.3: 625–34.

Jessica Hughes, 2018, 'The Texture of the Gift: Religious Teaching in the Greco-Roman World', in Graham Harvey and Jessica Hughes (eds), *Sensual Religion: Religion and the Five Eyes*, pp. 88–112, Sheffield: Equinox Publishing.

T. K. Johansen, 1997, *Aristotle on the Sense-Organs*, Cambridge: Cam-bridge University Press.

Danijela Kambaksovic and Charles T. Wolfe, 2016, 'The Senses in Philosophy and Science: From the Nobility of Sight to the Materialism of Touch', in Herman Roodenburg (ed.), *A Cultural History of the Senses in the Renaissance*, pp. 107–25, London: Bloomsbury Publish-ing.

Pieter J. Lalleman, 1998, 'Healing by a Mere Touch as a Christian Con-cept', *Tyndale Bulletin* 48.2: 355–61.

Louise J. Lawrence, 2013, *Sense and Stigma in the Gospels: Depictions of Sensory-Disabled Characters*, Oxford: Oxford University Press.

Joseph M. Magee, 2000, 'Sense Organs and the Activity of Sensation in Aristotle', *Phronesis* 45.4: 306–30.

Bruce J. Malina, 1981, *The New Testament World: Insights from Cul-tural Anthropology*, Atlanta: John Knox Press.

—— and Richard L. Rohrbaugh, 2003, *Social-Science Commentary on the Synoptic Gospels*, Minneapolis: Fortress Press.

Dale Martin, 2009, *The Corinthian Body*, Haven: Yale University Press.

Pascal Massie, 2013, 'Touching, Thinking, Being: The Sense of Touch in Aristotle's *De Anima* and Its Implications', *Minerva: An Internet Journal of Philosophy* 17: 74–101.

Filip Mattens, 2009, 'Perception, Body, and the Sense of Touch: Phenomenology and Philosophy of Mind', *Husserl Studies* 25: 97–120.

Monica Jyotsna Melanchthon and Mothy Varkey, 2020, 'Teaching Biblical Studies in a Pandemic: India', *JBL* 139.3: 613–18.

Roger S. Nam, 2020, 'Biblical Studies, Covid-19, and Our Response to Growing Inequality', *JBL* 139.3: 600–6.

Jacob Neusner, 1973, *The Idea of Purity in Ancient Judaism*, Leiden: Brill.

Jerome H. Neyrey, 1986, 'The Idea of Purity in Mark's Gospel', *Semeia* 35: 91–128.

Amélie Oksenberg Rorty and Martha C. Nussbaum (eds), 2003, *Essays on Aristotle's De Anima*, Oxford: Oxford University Press.

Baruch J. Schwartz, Naphtali S. Meshel, Jeffrey Stackert and David P. Wright (eds), 2008, *Perspectives on Purity and Purification in the Bible*, London: T & T Clark International.

Anthony Synnott, 1991, 'Puzzling Over the Senses: From Plato to Marx', in David Howes (ed.), *The Varieties of Sensory Experience: A Sourcebook in the Anthropology of the Senses*, pp. 61–78, Toronto: Toronto University Press.

Slavoj Žižek, 2020, *Panic!: Covid-19 Shakes the World*, New York: Polity Press.

3

Bodies in Covid:
A Caribbean Perspective

ANNA KASAFI PERKINS

This reflection begins with a quote from a pandemic poem
entitled 'Collage' by St Lucian poet John Robert Lee (2020):

To the saints who are in Ephesus

...
Who would think
that pestilence is ravaging our world?

No safe zone on continent or island,
familiar routines locked-down,
family, friends, lovers masked and distanced,
networks obsessed with flattening curves and death
    statistics,
churches and mosques closed, except for fanatics,
beaches, bars, brothels shut, except for the skeptics
or those who want normal here and now,
and there are us who must crowd long lines outside shops –

who wrote the script
who configured this incredible dystopia?

...
Judgement is dropping abroad
from our mouths and hands –

what unbelievable drama is rolling out behind the scenes,
Who is moving, Ephesians, to centre-stage of this cosmic
    scenario?

I juxtapose John Robert Lee's questioning prelude with an
excerpt from a 'Letter to the Editor' of the *Jamaica Observer*
(24 March 2020) titled 'Epidemiology, theology, and Covid-
19' by George S. Garwood:

> All mortals die. We are mortals; therefore, we die ... But, if
> some kind of life continues after our physical cessation this
> kind of life has to be what it always was – pre-existence.
> That is the absolute necessity for there to be the eternity of
> matter and energy – call it God if one wishes. If that, then, is
> the case, we can be consoled by the assurance that when 'we
> put off this mortal coil' we return to our eternal home – the
> cosmos.

Garwood attempts to provide a word of hope with a type of
cosmic theology that moves beyond the very 'mortal coil', that
is, the body and its attendant cloak of affliction, suffering and
death which elicited it. He wants to assure his readers that
there is life after death since matter and energy are eternal.

John Robert Lee, on the other hand, does not resile from
the concreteness of empty brothels and shut churches, even as
he faces the pandemic's cosmic, dystopian and eschatological
import, which perplexes 'the Ephesians'. Governments across the
world enforced precautions to 'flatten the curve' – quarantine,
lockdown, the isolation of infected individuals, travel restrictions, border closures, mask wearing and physical distancing
(Bochtis *et al.* 2020). These precautions have had and continue
to have a detrimental effect on the local and global economy
leading to significant economic crisis here and now.

At the same time, Lee, perhaps the foremost poet of the
Christian faith in the Caribbean, draws to our attention the
question of what is moving to centre-stage. His concreteness
centres 'our mortal coil', raising some implicit, if fundamental,
questions on the human experiences of bodies marked by the

pandemic. In particular, how are bodies treated in the Carib-bean, a region born out of and deeply marked by inequalities that divide, marginalize and exclude? The virus attacks not only individual physical bodies but also communal bodies, racialized bodies, gendered bodies, poor bodies – exploiting and deepening existing fault lines.[1] Simultaneously, our response to the pandemic exposes our deeper attitude towards the body. Importantly, for our consideration is the transformation of bodies that occurs in their (mis)treatment – the meanings become reinforced and in turn reinforce further practice (Bishop 2013).[2] Indeed, Covid reveals 'the management, manipulation, and pervasive political interventions into lives/deaths and (re)embodiments of not only "extremely" placed and exposed bodies, but including the "everyday" bodies of you and I' (Purnell 2020).

Autobiographical Pause

A Situated View

This reflection on the body in Covid is situated – it is not a view from nowhere. I am an African-descended/Black Atlantic Jamaican-born, Roman Catholic woman theologian, who dares to question the meaning of the body – 'this mortal coil' – as it has been laid bare(r) by the pandemic. My theo-ethical per-spectives have been shaped by womanist, postcolonial and critical race theory, which expose how bodies like mine are continually misrepresented and devalued by politico-religious systems, especially hegemonic Christianity.

Jamaican-born diaspora theologian Carol Marie Webster makes a two-fold argument about the Black Atlantic female body that provides a framework for viewing/critiquing the questions posed to the body in the Covid pandemic. Webster contends that 'Caribbean women's bio-political location and imaginings are deeply formed and informed the region's historical encounter with Christianity' (2017, p. 1). As such, under the Enslavement project, Black Atlantic bodies were *codified as innately inferior, destined for servitude and to be disposable/discard-able.* Black Atlantic bodies were 'ontolog-ically tarnished', and redemption was necessary and possible through sacralized violence perpetrated through the institution of chattel slavery and its norms that violated such bodies. Black Atlantic women in particular were positioned in Christian dis-course as not-human and not-woman and outside the frame of Christian moral regard. Black women's sexual, sensual, procreative, labouring, mothering potentials were co-opted, (mis)appropriated, owned, used, negated and abused. The power and impact of this religio-cultural violence on diaspora African women remain till today, often referenced in personal and cultural valuings and treatment based on shades of skin tone, body shapes, hair type, which impact the way we experi-ence the world (Chevannes 2014; Perkins 2011; Perkins 2017; Harris 2017).

Webster is insistent, however, that other scripts existed alongside this dominant violent and oppressive Christian one. Contending scripts arrived embodied in the African peoples brought as captives across the Atlantic; such scripts affirmed the bodies of African peoples, including women, as sacred and divine. Webster's project among Jamaican Catholic women unveiled/exposed the presence of such scripts among Roman Catholic women worshippers using the trope of Body as Temple. Webster reclaims these ways of being in the world that are not commonly articulated in the dominant discourse about African diaspora women. These ways prove resistant to the annihilation and devaluation of the person-body as strategies for flourishing and survival are interwoven into these bodily scripts.

It is such an analysis that forms the backdrop for my brief exploration of three actual situations from the pandemic in the Caribbean; these circumstances cause me to question the fundamental value given to bodies in Caribbean space, par-ticularly poor working bodies, 'loose' bodies, dead bodies. Of course, these three situations do not exhaust the challenge posed by Covid to the meaning of the body but simply serve to highlight the complexities of the situation. Importantly, these three circumstances highlight bodily vulnerability and the interpenetration of experiences; for example, 'loose' bodies and poor working bodies are also vulnerable migrant bodies, which may then be *codified as innately inferior, destined for servitude and to be disposable/discard-able* as dead bodies.

Religiously inf(l)ected

In addition, it is important to point out the necessity of this religiously infected-and-inflected exploration. There are implied theologies of the body/moral anthropologies which shape the way that a society treats the body. 'In other words, the ethics of the body is tied to the assumed metaphysics of the body' (Bishop 2013, p. 113). In the Caribbean, Christianity is/has been an important shaper of theologies of the body as well as a means of socialization into how to treat the body; however, it has an ambiguous impact as it both subordinates the body, especially the female body, while also empowering the body as a site of creativity and justice having been revalued by 'the Word that became flesh' (Perkins 2011). Of course, a faith that believes in a resurrected body or incarnated divinity does not necessarily value the body. As Schneider argues:

> The history of Christian thought [is] a gradual and inexorable erosion of the early Christian testimony of incarnate divinity. Over time and with the growth of an imperial, globally aspiring religion, actual bodies – of tax agents, fishermen, soldiers, prostitutes, carpenters, and children, for example- have been diminished in favour of immutable souls and the separation of divinity from world. (2010, p. 232)

'This Mortal Coil'

I take 'mortal coil' to mean the body (self, embodied spirit, ensouled body, human person). The expression is often mistaken as biblical but is actually Shakespearean, from *Hamlet*, where the main character frets in his famous soliloquy (Act 3, Scene I):

> For in that sleep of death what dreams may come,
> When we have shuffled off this mortal coil,
> Must give us pause – there's the respect
> That makes calamity of so long life.

Strictly speaking, 'mortal coil' refers to the troubles of life and the strife and suffering of the world (Rose 2020), which are perhaps captured in the biblical notion of 'the vale/valley of tears' (Psalm 84.6). It is used in the sense of a burden to be carried or abandoned. To 'shuffle off this mortal coil' is to die as mortal bodies do, as Garwood reminds us. Hamlet grieves the eventual and inevitability of death where the mortal coil is 'shuffled off'. Of course, the mortal coil has never been static as the meaning(s) of the body-which-is-naturally-marked-by-suffering-and-death has varied across time and place, shaped by important factors such as culture, religion, and even technology (Alberti 2016). This is no different in the Caribbean as we shall see below.

Dead Bodies

Some bodies are often the least countenanced, that is, bodies that have shuffled off their mortal coil and are the ultimate expression of our mortality – dead bodies. Different societies have different protocols for handling the dead. Most are focused on protecting the living from contamination by the dead (Alwala 2020). 'In many cases during emergency situations, dead bodies are no longer considered sacred and treated with utmost respect. On the contrary, they can even be a source of worry and fear' (Bayod 2020, p. 237). The fear and worry attending dead bodies during pandemics is compounded by the fact that cultural and religious rituals for preparing and sending the deceased to the afterlife, while allowing the living to grieve and celebrate, have been curtailed or even distanced during Covid (Sarmiento 2020).

In Jamaica, as in many other societies in the African diaspora, funerals and wakes have always been rituals of major significance (Alwala 2020; Paul 2007). Clinton Hutton maintains that during enslavement death and the rituals of transmigration 'became the cosmological roots of African-Caribbean freedom' (2019, p. 210). As a result, 'an entire universe of artistic methods and expressions and an aesthetic complex has developed around repossession and funerary rituals' (p. 210). These rituals of grief include 'nine nights' of wake-celebration and funeral rituals with often-elaborate coffins, fancily dressed mourners from home and abroad, carnival-like processions and performative tributes (Senior 1986; Imbert 1996).

> The coffin or casket in particular acquires the expressiveness of a status symbol signifying the importance both of the deceased and the bereaved family or community. Often disproportionate amounts of money are invested by poor communities in custom-made designer caskets. (Paul 2007, p. 143)

This is not unique to Jamaica. In a recent funeral in Trinidad, the deceased was driven across town sitting in the back of a pick-up; he was later buried in a casket (Steinbuch 2020). These 'bling bling' funeral rituals are more marked among the marginalized, whose practices differ from the 'resplendence of rites associated with the Rastafarian funerals … and the refined rituals of the middle class' (Paul, p. 152). Annie Paul (2007) argues that such practices show the use of the body in the cultural-politics of postcolonial Jamaica. The marginalized signal their power and the presence of a 'rebellious counter-society' by such displays. Not to be ignored, of course, is that these rituals have spiritual dimensions as they permit the community to make sense of their loss and pray for the deceased. Such practices are avenues for religious experience. Further, death and burial practices assist the bereaved to accept the death of their loved ones (Sarmiento 2020). It is for this reason that Roman Catholic Archbishop of Port of Spain in Trinidad, Jason Gordon, laments, 'Burying the dead is one of the works of mercy. And in this time of coronavirus, we are unable to do it in the way we are accustomed' (quoted in *Stabroek News* 2020).

Curtailed Funeral Rituals

No longer able to gather in numbers and mourn in the ways we are accustomed to; mourners are reduced to a few family members and the funeral home attendants; there is no viewing or touching or kissing of the departed. This is compounded further where families are not able to even bathe or dress their loved one at death. For Jamaicans, the bathing and dressing of the dead and the entire burial process are 'rituals of respect' – by which the dead transition to become ancestral spirits (Chevannes 2014, p. 18). In response to the curtailment of these rituals, some families refuse to bury their dead during the pandemic as they would not be satisfied with just a 'likl funeral':

> Nothing cannot be done, we cyah do nothing because of the coronavirus. Her children abroad and others will also like to come, so it affi push back, it wouldn't be fair ... We a pray to God this may pass over as quick as possible because she needs her proper send-off. (*Stabroek News* 2020)

The 'proper send-off' is part of paying respect to the dead (Chevannes 2014). Ironically, the first corona death in Jamaica was a woman who visited from the UK to attend a funeral. She infected several persons, eventually leading to the quarantine of one entire community (see Amante *et al.* 2020). Of course, there are numerous persons who, in refusing to give up their traditional practices, fall afoul of the law at home and abroad.

Persons have had to adapt and evolve new ways of honouring the dead, including Zoom funerals, livestreaming services, opting for cremation (Clarke 2020). The funeral industry has been affected. Customary 'old fashioned' church services with lengthy funeral programme (numerous tributes, remembrance and eulogy) are no more. Truncated ceremonies at the burial site are becoming the norm. Religious leaders have had to adapt to the circumstances and perform burial rituals that obey the law while providing the support that bereaved families needed. There may be scope for funeral and burial rituals to evolve to more ecologically friendly internments such as biodegradable coffins or shrouds at grave sites. In fact, the funeral home in Trinidad responsible for the deceased seated at his own funeral is contemplating aquamation – a water cremation.

No Longer Respected

Another troubling matter concerning the treatment of the covid dead cannot be ignored – the dead are no longer respected but are now feared and fearsome.[3] Initially, some Jamaican funeral homes refused to prepare them for burial, further compounding the grief of the families, many of whom had been stigmatized and shunned by their communities.[4] As told by one mortician contracted by hospitals to remove and store covid deceased: 'Another [mortician] was doing a removal for a funeral home. I asked him where are your PPEs? Him say, "Is what?" I said to him, "It's a Covid-19 body" and he just drove away – scared – he drove away' (Hibbert 2020). Some funeral home personnel treat the dead casually, without regard to the circumstances of their passing. These circumstances are exacerbated by the lack of regulation of Jamaica's mortuary industry and the absence of appropriate guidelines for handling covid deceased.

The World Health Organization has issued guidelines for persons suspected or confirmed to have died from Covid-19, including the preparation of the body, the wake and burial (WHO 2020). WHO notes that the possibility of transmission from dead bodies continues to be low; nonetheless, care must be taken to safeguard those handling these bodies. Importantly, WHO reiterates that 'the dignity of the dead, their cultural and religious traditions, and their families should be respected and protected throughout' (p. 1). At the same time, measures should be taken to respect the dignity of the dead. Family and friends are not barred from viewing the bodies of their dead after they have been prepared for burials, in accordance with local customs. However, they are not to touch or kiss the body, as is the local custom in Jamaica, and should perform hand hygiene after the viewing.

These realities raise numerous questions: What happens when our deeply human need to gather, grieve and console collides with the reality of a pandemic? What happens when the bodies of the dead become objects of fear and scorn, desecration even? What is clear is that people do not lose their dignity even in the face of death (Perkins and Landis 2020).

Working Bodies

The body is a medium through which capacity to work is possible. The body is 'the storehouse of labour power' in a neo-liberal context where production and consumption drive all (Lewis 2004, p. 243). Indeed, bodies cannot be conceived outside of the socio-economic and political context within which they operate. At the same time, bodies become commodified through their labour. Caribbean people are particularly sensitive to the commodification of bodies, given their history of enslavement where the bodies of their ancestors were deemed chattels and their labour, including reproductive labour, was forcibly appropriated by Europeans in plantation society (Webster 2017). Bodies, therefore, must be understood as 'a site of production of commodities' (Lewis 2004, p. 243) as well as a commodity consumed by others.

Perhaps no other bodies 'work' in the same way as those of sex workers who literally 'sell their bodies' to be 'consumed' by others. Sex workers, because of the nature of their work, are at increased risk at Covid time (see 'Loose Bodies' below). Other working bodies at increased risk during Covid, include migrant farm workers.

Farm Workers

Innumerable Caribbean people have lost their jobs and live-lihoods due to Covid. Among the most affected are migrant workers who ply their trade both inside and outside the region. Many Caribbeans undertake farm work in North America under the Seasonal Agricultural Workers Programme (SAWP). Emergency travel bans as well as high levels of infections and changed family circumstances of workers considerably decreased the seasonal workforce available to harvest crops in countries that rely heavily on seasonal workers. Given concerns with food supplies and supply chains, farm workers are often designated 'essential workers', yet they are generally not covered by national health schemes, they earn low or subsistence wages, have little to no savings, and may be debt-bonded. They are ineligible for state Covid relief; they cannot easily return home or find alternate employment abroad due to travel and visa restrictions (Kempadoo 2020). The precarious position of farm workers is not new. Before the pandemic, farm workers lived and worked in overcrowded and dangerous conditions. Now, given the circumstances, they are among the most vulnerable to covid infection (Bochtis 2020).

The Jamaican farm workers in Canada are a case in point. In April 2020, one such worker wrote an anonymous letter for fear of reprisal lamenting the circumstances under which they work; their situation was in stark contrast to the story told by their Canadian employer (Wilkins 2020). The Jamaican farm worker was at a property that was at the centre of a Covid outbreak and claimed to be forced to live with workers who had tested positive, sharing living quarters and utensils, with no provision of sanitizers or other protections. Among the affected Jamaicans were several who arrived in Canada after signing waivers that relieved the Jamaican government of any liability in the event they got sick or died as a result of the virus. Asking farm workers to sign such waivers seems to 'indicate that there is a greater concern by the Jamaican government for the state of the economy than the welfare of its people, with migrant workers being expected to carry some of the greatest [personal,] social and economic risks' (Kempadoo 2020). It is difficult not to see those workers as *codified as innately inferior, destined for servitude and to be disposable/discardable.* One such 'disposable and discardable body' lamented:

> This is our job, this is how we survive, this is how we take care of our family back home. Without this, God help [us ...] We are happy for it [jobs], but we need to be treated as equal as everyone ... We haven't seen nor hear from the [liaison officers ...] We want to feel comfortable working [so] that if we get injured we are treated equal. This could have been avoided ... When workers took sick, they took too long before [they got] medical attention and [were] still going to work, then it spread ... Please hear our cry. (Wilkins 2020)

Larry Lohmann (2020) would classify these farm workers as wage workers whose bodies constitute modern transnational capitalism. These workers are expected to show up on time every day, get only so many sick days each month, and make money for the boss year after year. Business is used to assuming that the health of these bodies is predictable, just as it has taken for granted that the unpaid work by the world's 'women, colonies and nature' will always be there to exploit.

Again, no bodies are simply disposable or fungible. Nor are they to be deliberately sacrificed for economic gain. However, with the large numbers of willing persons waiting to participate in the opportunities of farm work, there will always be more bodies available to sacrifice. The Jamaican government's responsibility to these workers should include the protection of their fundamental right to good health as well as their ability to work in order to contribute to their livelihood. Part of this should include negotiations so that these essential workers, who are contributing to the economic growth of both countries, participate in income support programmes and the Canadian government to mandate and monitor protocols to protect these essential workers.

'Loose' Bodies

The experience of sex workers during Covid is another site for reflection and interrogation. Already involved in an industry that is fraught with danger, moral opprobrium and discrimination, sex workers are portrayed as loose (immoral), criminal, vectors of disease and (willing?) victims of trafficking; they exist as 'unwanted forms of diversity' within the body politic. This is evident, for example, in the fact that 'sex work' and 'sex workers' are not legal terms in any Caribbean country. The stigmatizing and denigrating term 'prostitution' continues to be used to describe the activities of sex workers.

'Sex work' is as an income-generating activity, in most cases done informally and in an unregulated manner. Sex work in the Caribbean is diverse, covering a range of activities and settings including brothel, massage parlour, club, tourist-oriented, street-based, exotic dancing and escort services. Notably, sex work is a significant activity within the tourist industry. Some sex work activities and settings such as webcamming, telephone sex and live voyeurism are not a significant part of the trade in the Caribbean currently (Matolcsi *et al.* 2021).

Sex work in Christian-hegemonic spaces like the Caribbean are shaped by beliefs around its immoral/sinful status and the extent to which it undermines the God-given role for women in society. The church considers sex work to be sinful and against the divine will for sex, which is only moral in the context of heterosexual marriage. Feminists are also at odds on the issue as some consider sex work to be a manifestation of patriarchy and therefore a threat to the dignity and value of women; others regard it as legitimate work that requires legal protections. Clearly, any discussion around sex work revolves around 'morality' but it ought also to be about labour.

Caribbean countries deal with sex work in a prohibitive fashion, judging it to have negative implications for public morals and is therefore to be criminalized. The sex worker becomes the focus of punishment, but not the client or facilitators. This does not stop sex work but increases the vulnerability of sex workers to harassment by state agents, especially the police, and violence from clients and pimps.

With covid lockdowns and curfews, there has been a curtailment of the sex trade; sex workers are unprotected, increasingly vulnerable and unable to provide for themselves and their families. At the same time, according to Barbadian sex industry expert Charles Lewis, 'rising joblessness brought on by the pandemic has pushed more people to the sex trade for economic survival' (Henry 2021). Sex workers in Jamaica find it difficult to protect themselves (with sanitizers for themselves and their clients) or, given the nature of their job, to maintain social distancing (Williams 2020). The sex trade has become more competitive, prices have dropped for sexual services and the demand has shrunk (Henry 2021). Many Guyanese sex workers, like their counterparts across the region, have found it difficult to access the social support provided for formal-sector workers by the government. 'The problem is that many in authority don't see sex work as work' (UNAIDS 2020).

Sex workers face increased discrimination and harassment from agents of the state. Kempadoo (2020) details that the first person detained by the police in Barbados for breaking the curfew, introduced on 27 March, was a young Jamaican woman. She was scapegoated and made a public example of what could happen if curfew restrictions were breached:

> An easy target – young, a woman, alone, walking the street at night, poor and far from home – the typical image of the 'loose woman', and thus automatically deemed irresponsible, illegal and punishable. She was the target of the kind of sexual profiling that migrant working women must often deal with. (Kempadoo 2020)

In response, the Global Network of Sex Work Projects (NSWP) and UNAIDS draw attention to the hardships facing sex workers globally and call on nations 'to ensure the respect, protection and fulfilment of sex workers' human rights' (UNAIDS 2020). It is difficult not to see those workers as *codified as innately inferior, destined for servitude and to be disposable/*

discardable. Their experiences cry out for more complex and respectful ways of viewing labour in the Caribbean, especially sexual labour. Women (and increasingly men and transgender persons) engage in sexual-economic labour in a fashion that cannot be simply dismissed using the language of coercion or patriarchy. Critical examination is needed of the ideologies about women's sexuality, Christian-influenced or no, in order to interrogate and temper the moral indignation, stigma and criminalizing that currently surrounds sexual-economic activities (Kempadoo 2016).

Considerations and Conclusions

Pandemics test many of the moral perspectives that we take for granted. Covid seems to defy meaning as it impacts every dimension of our lives. Some may well argue that it is such a profound experience that it defies or lacks meaning (Meylahn 2020). Yet, Covid is laden with meaning in a fashion that questions the very meaning of humanity, the body, the mortal coil. Covid is loaded with meanings about the body, arising from and shaped by religious perspectives. Importantly, the pandemic highlights that the body is not simply passive, but also active, proactive and transformative. The body has the capacity to initiate change, offer resistance, irrespective of the environment in which it survives (Lewis 2004).

Covid, as with other public health crises, exposes existing meanings of the body – particularly of dead, working, 'loose' bodies – that reinforce the way they are (mis)treated. Attention is required for inequalities that disproportionately affect bodies that are desacralized, criminalized, marginalized and living outside of systems of respect, health and social protection. Living in a pandemic does not change the fundamental duty to respect and care for such bodies. Living in a pandemic does not change the requirement of justice for vulnerable and disadvantaged workers. The impact of Covid on these bodies highlights vulnerability as a moral concept:

Vulnerability captures the capriciousness of the human condition, or something like the concept of moral luck, but it also refers to a state of being which is created and intensified by unjust social conditions. By recognizing this distinction, we can discriminate between the vulnerability of human beings to suffering caused by a virus, on the one hand, and the particular injustice that groups like [sex workers, migrant workers, decedents] face during this pandemic, on the other. (Dunn 2020, p. 353)

The pandemic has a more devastating impact on such groups in the Caribbean because of the higher risk to their health and lives resulting from their working conditions, and also in terms of their socio-economic and political status and well-being (Madarova *et al.* 2020). Those workers are more vulnerable in the crisis because they are *codified as innately inferior, destined for servitude and to be disposable/discardable.*

Notes

1 Covid syndemic: Importantly, as the impact of the pandemic on bodies is contemplated and experienced, Horton's discussion in the *Lancet* proves insightful. What we have in this pandemic are two categories of disease interacting within specific populations: 1) an infection with severe acute respiratory syndrome (SARS-COV-2) and 2); an array of non-communicable diseases (NCDs) or what are commonly called underlying conditions or comorbidities.

These conditions are clustering within social groups according to patterns of inequality deeply embedded in our societies. The aggregation of these diseases on a background of social and economic disparity exacerbates the adverse effects of each separate disease. Covid is not a pandemic. It is a syndemic. The syndemic nature of the threat we face means that a more nuanced approach is needed if we are to protect the health of our communities (Horton 2020, p. 874).

Such a medical-social appraisal of the pandemic highlights the importance of attention to the social in what is too often treated as simply an epidemiological phenomenon.

2 At the heart of the discussion is the recognition that Covid is not the first pandemic to affect the region; for the last 40 years, the region has been living under the shadow of HIV and AIDS and several others

that have been transient if lingering. Even more fundamental to the discussion is Kiple's (1984) contention that disease played a profound role in the history of the Caribbean, which bears some reflection.

3 The bodies of the Covid dead have also been desecrated, as in one reported case of necrophilia in Guyana (*Jamaica Gleaner* 2020).

4 Some undertakers have complained that they had been given Covid victims to prepare for burial without being told the cause of death. And families complained of being rejected and stigmatized when a loved one died.

References

Fay Bound Alberti, 2016, *This Mortal Coil: The Human Body in History and Culture*, Oxford: Oxford University Press.

Bernard Alwala, 2020, 'The Fate of Prosperity Gospel in Kenya', *East African Journal of Traditions, Culture and Religion* 2.1 (2020): 23–33 (doi:10.37284/eajtcr.2.1.152).

Angelo Amante, Parisa Hafezi and Hayoung Choi, 2020, '"There are no funerals": Death in quarantine leaves nowhere to grieve' (accessed 16.3.21: www.reuters.com/article/us-health-coronavirus-rites-insight/there-are-no-funerals-death-in-quarantine-leaves-nowhere-to-grieve-idUSKBN2161ZM).

Rogelio P. Bayod, 2020, 'Covid-19 Age: Spirituality and Meaning Making in the Face of Trauma, Grief and Deaths', *Eubios Journal of Asian and International Bioethics* 30.5 (June): 237–42.

Jeffrey P. Bishop, 2013, 'Body Work and the Work of the Body', *Journal of Moral Theology* 2.1: 113–31.

Jack Black, 2020, 'Covid-19: Approaching the In-Human', *Contours: Journal of the SFU Humanities Institute* 10:1–10.

Dionysis Bochtis, Lefteris Benos, Maria Lampridi, Vasso Marinoudi, Simon Pearson and Claus G. Sørensen, 2020, 'Agricultural Workforce Crisis in Light of the Covid-19 Pandemic', *Sustainability* 12, 8212 (doi:10.3390/su12198212).

Barry Chevannes, 2014, 'Rastafari and the Coming of Age: The Routinization of the Rastafari Movement in Jamaica', in Michael Barnett (ed.), *Rastafari in the New Millennium: a Rastafari reader*, pp. 13–32, Syracuse, NY: Syracuse University Press.

Paul Clarke, 2020, 'New Signs of Life Among Dead – Funeral Home Operators Hopeful For Full Revival As Covid-19 Restrictions Ease', *Jamaica Daily Gleaner*, 14 July.

Shannon Dunn, 2020, 'Covid-19 and Religious Ethics', in Toni Alimi, Elizabeth L. Antus, Alda Balthrop-Lewis *et al.*, 'Covid-19 and Religious Ethics', *Journal of Religious Ethics* 48: 349–87.

Jewel Gausman, and Ana Langer, 2020, 'Sex and Gender Disparities in the Covid-19 Pandemic', *Journal of Women's Wealth* 29.4: 465–6 (doi: 10.1089/jwh.2020.8472).

Niels Henrik Gregersen, 2020, 'The corona crisis unmasks prevailing social ideologies', *Dialog* 1–3.

Dawn P. Harris, 2017, *Punishing the Black Body: Marking Social and Racial Structures in Barbados and Jamaica*, Athens: University of Georgia Press.

Anesta Henry, 2021, 'Curfew "upsetting" sex workers', *Barbados Today*, 6 January (accessed 16.3.21: https://barbadostoday.bb/2021/01/06/curfew-upsetting-sex-workers/).

Kimberley Hibbert, 2020, 'Covid-19 slackness concerns funeral directors', *Jamaica Observer*, 4 October (accessed 16.3.21: www.jamaicaobserver.com/news/covid-19-slackness-concerns-funeral-directors_204549?profile=1373).

Richard Horton, 2020, 'Offline: Covid 19 is not a pandemic', *The Lancet* 396, 26 September, 874 (www.thelancet.com).

Clinton Hutton, 2019, 'Introduction: African-Caribbean Spirituality and Creativity', *Caribbean Quarterly*, June: 207–11.

Maura Imbert, 1996, 'Who Go Bury She?', *Pastoral Bulletin* 12.1 (October): 25–9.

Jamaica Gleaner, 2020, 'Guyanese Man Gets 3 Years for Having Sex with Dead Covid-positive Woman', 1 October (accessed 16.3.21: http://jamaica-gleaner.com/article/caribbean/20201001/guyanese-man-gets-3-years-having-sex-dead-covid-positive-woman).

Kamala Kempadoo, 2016, 'The War on Humans: Anti-trafficking in the Caribbean', *Social and Economic Studies* 65.4: 5–32.

———, 2020, 'On Migration, Sex Work and the Pandemic in the Caribbean', *Stabroek News*, 4 May (accessed 16.3.21: www.stabroeknews.com/2020/05/04/features/in-the-diaspora/on-migration-sex-work-and-the-pandemic-in-the-caribbean/).

Kenneth F. Kiple, 1984, *The Caribbean Slave: A Biological History*, Cambridge: Cambridge University Press.

John Robert Lee, 2020, *Pierrot*, Leeds: Peepal Tree Press.

Linden Lewis, 2004, 'Masculinity, the Political Economy of the Body, Patriarchal Power in the Caribbean', in Barbara Bailey and Elsa Leo-Rhynie (eds), *Gender in the 21st Century: Caribbean Perspectives, Visions and Possibilities*, pp. 236–61, Kingston: Ian Randle.

Larry Lohmann, 2020, 'Covid-19 and the End of the Modern Working Body', *The Corner House*, July (accessed 16.3.21: www.researchgate.net/publication/342923865_Covid-19_and_the_End_of_the_Modern_Working_Body).

Zuzana Madarova, Pavol Hardos and Alexandra Ostertagova, 2020, 'What Makes Life Grievable? Discursive Distribution of Vulnerability in the Pandemic', *Journal of International Relations* 55.4: 11–30.

Andrea Matolcsi, Natasha Mulvihill, Sarah-Jane Lilley-Walker, Alba Lanau and Marianne Hester, 2021, 'The Current Landscape of Prostitution and Sex Work in England and Wales', *Sexuality & Culture* 25: 39–57 (doi:10.1007/s12119-020-09756-y).

J-A. Meylahn, 2020, 'Being human in the time of Covid-19', *HTS Teologiese Studies/Theological Studies* 76.1: a6029 (doi:10.4102/hts. v76i1.6029).

Annie Paul, 2007, '"No Grave Cannot Hold My Body Down": Rituals of Death and Burial in Postcolonial Jamaica', *Small Axe* 11.2: 142–62.

Anna Kasafi Perkins, 2009, '"God (Not) Gwine Sin Yuh": The Female Face of HIV/AIDS in the Caribbean and a Theology of Suffering', in Mary Jo Izzio and Mary Doyle Roche (eds), *Calling for Justice throughout the World: Catholic Woman Theologians and the HIV-AIDS Pandemic*, pp. 84–96, Continuum.

———, 2011, 'Carne Vale (Goodbye to Flesh?): Caribbean Carnival, Notions of the Flesh and Christian Ambivalence about the Body', *Sexuality & Culture*, August, pp. 361–74.

———, 2017, 'Shi Wi Use Har Blood Tie Him': A Theological Interrogation of Cultural Beliefs about Menstruation and Female [Im] morality in Jamaica', in Nicholas Faraclas *et al.* (eds), *Memories of Caribbean Futures: Reclaiming the precolonial to reimagine a postcolonial languages, literatures and cultures of the Greater Caribbean and beyond*, pp. 349–59, Puerto Rico/Curacao: University of Curacao.

———, 2019, 'Christian Norms and Intimate Male Partner Violence: Lessons from a Jamaica Women's Health Survey', in Antipas L. Harris and Michael D. Palmer (eds), *The Holy Spirit and Social Justice Interdisciplinary Global Perspectives: History, Race & Culture*, pp. 240–67, Lanham, MD: Seymour Press.

——— and Clive R. Landis, 2020, *Ethics Amidst Covid: A Brief Handbook for Policy Makers in the Caribbean*, Smashwords Edition.

Kandida Purnell, 2020, 'The Body Politics of Covid-19', *The Disorder of Things*, 6 April (accessed 16.3.21: https://thedisorderofthings. com/2020/04/06/the-body-politics-of-covid-19/#more-17655).

Jennifer Rose, 2020, 'The Mortal Coil of Covid-19, Fake News, and Negative Epistemic Postdigital Inculcation', *Postdigital Science and Education* 2: 812–29.

Philip Joseph D. Sarmiento, 2020, 'Changing Landscapes of Death and Burial Practices', *Journal of Public Health* (Correspondence): 1–2.

Laurel C. Schneider, 2010, 'Promiscuous Incarnation', in Margaret D. Kamitsuka (ed.), *The Embrace of Eros: Bodies, Desires, and Sexuality in Christianity*, pp. 231–46, Minneapolis: Fortress Press.

Olive Senior, 1986, 'Country of the One-Eye God' in *Summer Lightning and other stories*, Harlow: Longman.

William Shakespeare, 'Speech: "To be, or not to be, that is the question"', *Poetry Foundation* (accessed 16.3.21: www.poetryfoundation. org/poems/56965/speech-to-be-or-not-to-be-that-is-the-question).

Stabroek News, 2020, 'Trinidad: '5-minute funeral' for Covid-19 victim Miss Vernise', 2 April (accessed 16.3.21: www.stabroeknews.com/ 2020/04/02/news/regional/trinidad/trinidad-5-minute-funeral-for-covid-19-victim-miss-vernise/).

Yaron Steinbuch, 2020, 'Dead Man Banned from his Own Funeral after Arriving on Chair', *New York Post*, 3 December (accessed 16.3.21: https://nypost.com/2020/12/03/dead-man-banned-from-his-own-funeral-after-arriving-on-chair/).

UNAIDS, 2020, 'Guyana community organization serves sex workers on the edge during COVID-19' (accessed 31.3.21: www.unaids.org/ en/resources/presscentre/featurestories/2020/july/20200729_Guyana_ SW_coalition)

Carol Marie Webster, 2017, 'Body as Temple: Jamaican Catholic Women and the Liturgy of the Eucharist', *Black Theology: An International Journal* 15: 1–20.

Arlene Wilkins, 2020, '"Hear our cry": J'can worker on Covid-hit Canadian farm appeals for assistance', *Jamaica Observer*, 30 April (accessed 16.3.21: www.jamaicaobserver.com/front-page/-hear-our-cry-j-can-worker-on-covid-hit-canadian-farm-appeals-for-assistance_ 193151?profile=1606).

Andre Williams, 2020, 'Covid cleanness hard in a dirty business', *Jamaica Daily Gleaner*, July 6 (http://jamaica-gleaner.com/article/ lead-stories/20200706/covid-cleanness-hard-dirty-business).

World Health Organization (WHO), 2020, 'Infection prevention and control for the safe management of a dead body in the context of Covid-19', *Interim guidance* (4 September).

4

Spiritualities in Resistance: Latin-American Social Movements and Solidarity Actions

ANGELICA TOSTES AND

DELANA CORAZZA

When most people think of spirituality, they imagine a great divinity – something superior, outside of and in addition to ourselves. We on the other hand think that it is a mistake to confine spirituality to the great and luminous experiences in the divine-human relationship. We are thus in step with *Metal e Sonho* by Pedro Tierra (2013), which calls upon colleagues to 'organize hope' (*Organizar a esperança*) and 'lead the storm' (*Conduzir a tempestade*) that will create the 'world of freedom' (*um mundo de liberdade*). For Tierra, we do not need permission to break through the walls of the night to reach this new world of freedom.

Spirituality relates to the demands of our daily lives; it is about finding hope in the midst of chaos. This hope is not in the promise of future paradise or some eschaton, but it is based in concrete and touchable things. The sacred and the profane mix in everyday life. The cup of hot coffee, the daughter or son who comes home alive after work, the potable water to drink, a roof to live in, the music that resonates through small churches in impoverished communities ... in those we experience spirituality. As Ivone Gebara suggests, 'the daily life is the life of everyday with its repetition, differences and improvises. It is on life of everyday that we seek the everyday bread and that we suffered violences, aggressions, worries just like the small joy and surprise that came unexpectedly' (p. 157).

Buen Vivir

Spirituality as a path to resistance is beyond religion, creeds or temples. Rather, it is a spirituality we find in the streets, in marches, in the screams seeking justice and in the gentle hands that help others. Our theoretical reference is the decolonial perspective of the Quechua concept *of sumak kawsay* (*buen vivir*, good living or living well).

Many Latin American indigenous cultures have revalued the *Buen Vivir* proposal from their ancestral traditions. In the Andes, the Quechua speak of *Sumak-wasay* as living well, (quality life). The Aymara speak of *Sumak-Kamana* (the good life). The Guarani speak of *nhandereku* and *teko-porã* (living life for real and in a community way – well-being is living well) (Barros 2018).

Buen vivir is a wide concept, not well translatable in English. It is not individual the well-being promoted by capitalism. Far from it. *Buen vivir* is a community-centric idea of harmony: with humans and nature. The clutches of imperialism and capitalism continue to threaten *buen vivir*, and the coronavirus exposed the weaknesses of a society based on neoliberal and egoistic principles.

> The Buen Vivir has reference to live in harmony, in balance, to decide, respecting and enhancing differences, diversity, along with complementarities. It is also about beautiful living, it implies a strained relationship with nature, which is not conceived as an untapped bank of resources, like the Pachamama, Mother Earth with which is in an indispensable relationship. There, life flourishes with beauty and prodigality, there is growth in food, which in turn requires care from the people, the respect, the attention, the collective work, the minga. In relation to the Pachamama, the cultivation and the life of the animals live, as well as the dance and the party,

all suppose with respect to natural cycles, and the sacrality is present in continuous form. Work, worship and festival are inseparable. (Ramos 2015, p. 214)

Coronashock

Latin America must be understood not only as a geographic space, but first as a socio-political, cultural and epistemic territory forged by and under colonialism, but with a potential for decolonization (Mignolo 2003). Colonialism ended with independence, while 'coloniality' refers to the colonial logic, thought and legacy that survived. This legacy is present in the structures of society and institutions, in imaginary, mind, subjectivities and epistemologies until the present (Castro-Gomez and Grosfoguel 2007).

The territory of Abya Yala (Latin America) was invaded in 1492 by Europeans; before that, an estimated population of between 57 and 90 million identified themselves as '*maia, kuna, chibcha, mixteca, zapoteca, ashuar, huaraoni, guarani, tupinikin, kaiapó, aymara, ashaninka, kaxinawa, tikuna, terena, quéchua, karajás, krenak, araucanos/mapuche, yanomami, xavante* between so many other nationalities and native people in this continent' (Porto-Gonçalves 2009, p. 26). All those people had their own spiritual practices, beyond the concept of religion and the duality of culture and religion. Our languages, cultures and existence were colonized and understood as inferior. We suffered the complete extermination of entire populations, slavery, the (dis)possession of lands, exploitation of nature's riches and sexual violence (Porto-Gonçalves and Quental 2012).

This legacy is present in the structures of society and institutions, in imaginations, and in minds, subjectivities and epistemologies even today (Castro-Gomez and Grosfoguel), and it has a name: coloniality. Coloniality is a concept that was introduced by Peruvian sociologist Anibal Quijano, in the 1980–90s, and expounded by Walter Mignolo in his book *Local histories/global designs: Coloniality, subaltern knowledges, and border thinking* (2012). Grosfoguel (2010) argues that the concept of coloniality allows us to understand the colonial forms of domination after the colonial administration, made by the colonial cultures and by the world-capitalist-modern-colonial system.

In the spectrum of coloniality, Latin America is struggling with coronashock. It is important to understand that corona-shock is a term that refers to how a virus struck the world with such gripping force; it refers to how the social order in the bourgeois state crumbled, while the social order in the social-ist parts of the world appeared more resilient (*Tricontinental* 2020a). In *Tricontinental* we can see what happened in Latin America under coronashock:

> the [Covid] pandemic has furthered – sometimes dramatic-ally – a series of economic and social processes that were already underway before the virus emerged. Capitalism's crisis of legitimacy ... also has put on display the resound-ing failure of neoliberal policies to effectively combat the health and social crises. Finally, the current situation puts into question the effects, actions, and challenges that these processes pose for people's movements and the alternatives that they are creating. (*Tricontinental* 2020b)

What we have heard since the pandemic started is that the fear of hunger was greater than the fear of the virus. But the hunger of poor communities was not a key concern for political leaders who talk about restarting the economy. In a short time, the residents of the peripheries of our cities directly felt what it means not to be able to work. Staying at home for the major-ity of the working-class can mean the impossibility of survival due to the lack of economic alternatives, or due to the violence of the State (especially in the peripheral territories, where it murders some of the population), or due to the domestic vio-lence suffered by women who, without alternatives, have had to live 24 hours a day with their abusers. With no way to sup-port themselves and their families, with difficulties in accessing emergency aid (which is deposited monthly for families that could prove low income), residents of the peripheries of the cities had to rely on solidarity to survive.

Brazil is facing an oppressive and neo-fascist government and has become the regional epicentre of Covid and one of the centres of the pandemic on a global level. 'The failure of the federal government to adopt sufficient measures to combat the pandemic has created a catastrophic situation that is edging towards a humanitarian tragedy. This inaction by the federal government is accompanied by President Jair Bolsonaro's approach of underestimating, or even denying the problem, and placing concern for the economy before concern for the people' (*Tricontinental* 2020b). Solidarity actions are the way to rethinking the world we want. As Arundhati Roy wrote: 'pandemics have forced humans to break with the past and imagine their world anew. This one is no different. It is a portal, a gateway between one world and the next' (2020).

Periferia Viva (Living periphery): Solidarity as Resistance

We focus on the Brazilian context and how social movements are challenged by coronashock. In a historical connection between the countryside and the city, several movements target hunger. Agrarian reform has given concrete answers to overcome the health crisis that has affected the workers of our country in different ways. Food crops planted and harvested without poison, the result of family farming and organized peasants, have become a rich and abundant source of food in various corners of the country.

Several popular organizations – such as the Landless Rural Workers' Movement (MST), the Rights Workers' Movement (MTD), the Consulta Popular (Popular Consultation) and the Levante Popular da Juventude (Popular Youth Uprising) – are already operating in several of these territories in which hunger is fought. They get together and build a solidarity campaign called *Periferia Viva*. The donation of healthy food to the peripheries of the country, whether in the form of baskets or the preparation of lunch boxes, has been the milestone of the campaign and has enabled the building of bonds and organization in the territories.

To satisfy hunger, something so genuinely human was the putting of solidarity into practice and building processes of participation and popular organization. The campaign relied not only on social movements organized for the donation and distribution of food but also with the people who, upon receiving donations, felt part of the process and could also contribute, both helping in the distribution and in the mapping of the most vulnerable families and promoting exchanges of necessary items – such as warm clothing, wheelchairs, mattresses, among other things. They also participate through donations that activists have advanced in what we call the Battle of Ideas. It was by approaching this genuine need that the militants, in close dialogue with reality, were able to materialize the daily dispute of narratives, deeply believing that the recipients are able to see this reality and intervene in it.

In the course of the actions and in the advancement of the organization in the territories, new subjects are placed as a fundamental part of the process, advanced in the counter-narrative of the current president's denial of the virus. *The people taking care of the people* announces a new form of organization and being in the territory, their daily actions made it possible even more, to refine the look at the reality of these spaces. The opportunity to create a new militant from the practice of care is one of the elements of advancing the dispute for solidarity.

The question of the hungry people – 'Why are these landless people giving us food?' – has been important because of the innumerable possibilities. The militants were attentive and, as they told us, they continue to build the pedagogy of listening: listening to questions, understanding the reading of the world through them, 'dismantling the magical vision' of the 'will to God' as responses to the ills experienced, and building the answers together. Listening pedagogy also asks: 'Why are you hungry?' The answers transcend fatalism, generate new reflections and, necessarily, actions (see also Chapter 14 by Sithembiso Zwane).

Popular Solidarity: Spirituality beyond Religion

Popular solidarity is built with many hands. It is the impulse that leads us to act on behalf of our members by offering what we have, not what we have left. It is the people giving objective answers to the people that is the primary and fundamental way to delve into reality. Why do we have to fight? Why are we the ones who have to activate our basic rights to essentials such as land, food and security? Advancing the understanding of these questions *from practice* is a way to enhance transformation processes. Transforming from the popular organization is the seed for a new society.

The spirituality of well-being is beyond the religions and religiosities organized by the West. For us in Abya Yala, it is necessary to decolonize spirituality in order to bring spirituality to our experience in daily life and struggle. Spirituality is more than high theological discussions and more about giving meaning for life, finding strength to face injustice, poverty, neoliberalism, racism, machismo and the consequences of colonization.[1] For Gustavo Gutiérrez, 'Latin American spirituality is clearly characterized by a constant fundamental reference to reality ... It is a double reference, both of origin and purpose. Starting from and returning to reality: that is the unmagical realism of Latin American spirituality' (1989, p. 60).

In the period of the pandemic, militants of social movements have been putting themselves at risk, dealing with a government that is concerned with the economy of large companies and not with the lives of the working class and minorities. 'Solidarity is a risk,' said Lucas Lemos, social activist of Landless Workers' Movement. Spirituality is to put your own life at risk to save and help others.

For the construction of 'land without evils', the militants take in their backpacks a fine reading of reality, which overcomes the limited view that satiating hunger is something of assistance, a task of conservative institutions. No. This backpack is loaded with deep love for another human being that they see as themselves. It is because we like so many people in our country that this militant's backpack is filled with hope, not with waiting, but with concrete action, as our companion would say 'we learn things by doing'. It is in doing that hope is created. This backpack is also loaded with so many theoretical references, books of poetry, so many songs, but also so many teachings of other militants who, like them, at other times in history, were on the march in the struggle and construction for a new society. This backpack is also loaded with the hope of our people, who wake up at dawn to make coffee while donations are being organized. These many Marias that populate the territories of solidarity make their backpacks so much lighter. So many flags, so many dreams are in these backpacks!

Periferia Viva is a resistance spirituality movement, seeking to transform the world. Latin American liberation theology has always sought the divine in the most vulnerable, and through social transformation it builds the horizon of *buen vivir*. It offers decolonizing solutions in all areas of life. *Buen vivir* overcomes Cartesian dichotomies, interweaves linear time with circular time, myth with history, and the objectivity of production with the subjectivity of 'mother earth'. For Paulo Suess, the construction of *Buen Vivir* means (i) decolonizing political institutions, (ii) de-commodifying knowledge, faith, school, health and (iii) deprivatizing what should be in the public domain (Suess 2010).

The many peripheries of our country, flooded with stories and memories, have been the fertile land of a people who, used to helping and being helped in their daily lives, are now challenged to organize themselves and to understand their territory as a space of collective care and self-care. This popular solidarity, so present in the immediate need of so many people, is food for the struggle and the seed of progress in the argument for revolutionary solidarity. Senhorinha, an activist from Pernambuco, deepens the reflection (in conversation with the authors):

> Popular solidarity is the creative capacity that our people have, in times of greatest crisis, to collectively find the answers to this crisis. Our people already have the germ of solidarity: a mother takes her neighbour's son at school; Dona Maria who only has a kilo of beans, shares with José that has none;

but class solidarity is that we perceive ourselves as a people, as a class that is exploited, oppressed all the time – it is this need to help others by calling them the reality that is inserted. It is not teaching how to fish but asking why even though he may learn to fish, he does not have the rod.

It is already known that the capitalist system is incapable of producing the *buen vivir* of all citizens. Consumerism and hunger are expressions of imbalance in the distribution of land goods. Popular solidarity is opposed to neoliberal logic and seeks to create strategies and tactics to overcome and emancipate everyone. Ademar Bogo, Leonardo Boff and Frei Betto sought to systematize solidarity as a conscious action by people of the same class in the search for joint alternatives for definitive solutions for all:

> The value of solidarity. More than ever, solidarity becomes a fundamental value, but we must understand and develop it based on our class interests, within our territory and outside it … Among those included, solidarity has the character of 'collaboration'. When it comes from the included to the excluded, it has the character of 'assistance' … Solidarity represents attitudes completely opposite to collaboration. It must be the conscious action of people of the same class in the search for joint alternatives to seek definitive solutions for all. (Boff *et al.* 2005)

Popular solidarity is classist. Class character is the people who set themselves in motion towards their emancipation. Considering 14 million unemployed people or 40 per cent among unemployed, discouraged and beneficiaries of social aid, there is no prospect of a revolutionary process without involving them, without the protagonism of the working and impoverished class.

Tricontinental interviewed Kelli Mafort, from the national board of the Landless Workers' Movement, and she coined the term 'Solidarity Inc.', which works like charity – 'it is vertical, based on a relationship between those who have and choose to give and those in need who can only receive. People who receive donations are seen as nothing but recipients of donators ... is similar to what Paulo Freire called "the banking model of education"' (*Tricontinental* 2020c; Freire 1996, p. 57). Mafort further explains: 'Popular solidarity, on the other hand, arises within working-class communities; it is based on mutual help and respect and produces organizations that enhance people's dignity.'

To Inspire Us

Spirituality resides in risk, in solidarity that emancipates lives, that decolonizes the mind, the body, and that dignifies life. In addition to religions, faith-based people and communities must learn and drink from social movements, join in the struggles and constructions for *buen vivir*.

In 2020 we faced coronashock. In 2021 we are still dealing with the psychosocial and political wounds and scars of the virus and social injustices. We must not be discouraged, but rather, keep walking in the hope of horizons of justice, of democracy, and in the struggle so that no pot is empty, that no body is shattered by the violence of gender, race, ethnicity, class, sexuality. Our reflection ends with a quote from theologian Leonardo Boff (2009):

> Solidarity manifests itself in compassion (putting oneself in the other person's shoes), in affection, in the right partnership and in unconditional love for the oppressed class to be realized. It is best expressed in the free delivery of what is best, including life itself, for people and peoples to realize the eternal dream of universal brotherhood. If you feel the pain of others as your pain, if the injustice in the body of the oppressed is the injustice that hurts your own skin, if the tear that falls from your desperate face is the tear that you also shed, if the dream of the disinherited of this cruel and merciless society is your dream of a promised land, then you will be a revolutionary, you will have lived the essential solidarity.

Note

1 Machismo is a concept that can describe many aspects of Latin American male behaviour. It is not only defined as sexism or misogyny, but also the idea that men are superior to women. 'Machismo reinforces the idea of women as second-class citizens whose rights and opportunities – even when included in public policies – are undermined in their households, in the streets, at school or work' (Ortiz 2018).

References

Marcelo Barros, 2018, 'Educação e Espiritualidade nos caminhos do Bem-viver', *Revista Senso* (accessed 17.3.21: https://revistasenso.com.br/bem-viver/educacao-e-espiritualidade-nos-caminhos-bem-viver/).

Leonardo Boff, 2009 (Outubro), Caderno de Formação 38, Setor de Formação do MST, Método do Trabalho de base e organização popular.

———, Frei Betto and Ademar Bogo, 2005 (Outubro), Cartilha n. 09 da Consulta Popular, Valores de uma prática militante.

Santiago Castro-Gomez and Ramón Grosfoguel (eds), 2007, *El Giro Decolonial. Reflexiones para una diversidad epistémica más allá del capitalismo global*, Bogotá: Universidad Javeriana-Instituto Pensar, Universidad Central-IESCO, Siglo del Hombre.

Paulo Freire, 1996, *Pedagogia do Oprimido*, São Paulo: Paz e Terra.

Ivone Gebara, 2017, *Mulheres, religião e poder: ensaios feministas*, Edições Terceira Via.

Ramón Grosfoguel, 2010, 'Para descolonizar os estudos de economia política e os estudos pós-coloniais: transmodernidade, pensamento de fronteira e colonialidade global', in Boaventura de Souza Santos, Maria Paula Menezes (eds), *Epistemologias do Sul*, pp. 115–47, Coimbra, Portugal: Cortez.

Gustavo Gutiérrez, 1989, *Beber en su propio pozo*, 6th edition, Salamanca: Ediciones Sígueme.

Walter Mignolo, 2003, *Historias locales-diseños globales: colonialidad, conocimientos subalternos y pensamiento fronterizo*, Madrid: Akal.

———, 2012, *Local Histories/Global Designs: Coloniality, Subaltern Knowledges, and Border Thinking*, Princeton, NJ: Princeton University Press.

Veronica Ortiz, 2018, 'The Culture of Machismo in Mexico Harms Women' (accessed 17.3.21: https://merionwest.com/2018/01/28/the-culture-of-machismo-in-mexico-harms-women/).

Carlos Walter Porto-Gonçalves, 2009, 'Entre América e Abya Yala-tensões de territorialidades', *Desenvolvimento e Meio Ambiente* 20: 23–30.

———— and Pedro de Araújo Quental, 2012, 'Colonialidade do poder e os desafios da integração regional na América Latina', *Polis. Revista Latinoamericana* 31: 1–33.

Rosa Ramos, 2015, *Sumak Kawsay, suma Qamaña, tekopora vida buena. Una propuesta de la sabiduría indígena*, En M. Trejo y R. Hermano (org.): La reforma de la iglesia en tiempos de discernimiento, Montevideo: Revista Misión.

Arundhati Roy, 2020, 'The pandemic is a portal', *Financial Times* (accessed 17.3.21: www.ft.com/content/10d8f5e8-74eb-11ea-95fe-fcd274e920ca).

Paulo Suess, 2010, 'Elementos para a busca do bem viver (sumak kawsay) para todos e sempre', Conselho Indigenista Missionário (accessed 17.3.21: cimi.org.br/2010/12/elementos-para-a-busca-do-bem-viver-sumak-kawsay-para-todos-e-sempre/).

Pedro Tierra, 2013, *A Palavra Contra o Muro*, São Paulo: Editora, Geração.

Tricontinental, 2020a. 'CoronaShock and Socialism', *Tricontinental*, 8 July (accessed 17.3.21: www.thetricontinental.org/studies-3-corona shock-and-socialism/).

————, 2020b, 'Latin America Under CoronaShock: Social Crisis, Neoliberal Failure, and People's Alternative', *Tricontinental*, 7 July (accessed 17.3.21: www.thetricontinental.org/dossier-30-corona shock-in-latin-america/).

————, 2020c, 'Youth in Brazil's Peripheries in the Era of Corona-Shock', *Tricontinental*, 5 October (accessed 17.3.21: www.thetriconti nental.org/dossier-33-brazil-youth/).

5

intact

life in fullness
life of flourishing
wat a gwan
in these days

touch is absent
fear chases away relationships
pandemics rendering all untouchable
un-touch the other and live
un-touch the other resisting perceived dangers

dead bodies
having lost their mortal coil
having lost their respectability
remind of the sacredness of an afterlife Unknown
remind of the Creator and Ancestors long gone

touch not handle not
some rendered unclean
desires for purity transcending empathy
no care for the suffering nor for the dying
un-touch the other and feel healed

years of exploitation a pandemic of great proportion
economies once again a priority over living bodies
hunger and poverty abound in great proportion
multiplied exponentially
resulting in fire and rage in the streets

mortality is fragile
distance providing a mask of safety
amidst a drastic scenario
of mortal coils falling rapidly
side-stepped by societies isolating those labelled different

touching the sick
breaking down barriers
we reach out defying historic boundaries and borders
the hands-off perpetuation of patriarchy itself a death knell
the touch a resistance to this living hell

at center
are the exclusions
the attack on racialized and poor bodies
an exploitation of fault lines uncovered in our midst
workers exploited by neo-colonialism falling to social diseases
    named pandemics

bodies lying in state
bodies devalued
bodies laid bare
by pandemics of racism and economic distress
left unchecked

vulnerable lives
deemed as working bodies
display invisibility – a concept of social dis-ease
dead bodies walking above ground being ignored
lives of women and men in permanent economic quarantine

contaminated by living poor bodies left to rot
turned away by a system
spiritually bereft of compassionate response
nobody cares for lives that don't matter
necro-politics dictating who lives and who dies

un-mask the virtual realities
open closed eyes to the unholy illusions
of masked inequalities
of economies flourishing as they exploit the untouchable
of healing that can come in absence of touch

healing requires radical living
touch now bearing new meaning
touching across the miles
touching in new ways
touching to dismantle systems that isolate and reproach

resist the politicization of bodies
resist the pollution of lives made pure by Creator
resist the touch that violates
resist the touch that brings hurt
resist the narrow-minded thinking of touch
embrace new paths to touch and contact that heal

bodies are embodied by Divine image
mortal coils connected to Spirit without touching
feel the interconnectedness connecting all things living
moving us to healing the created
we wait for the healing

we wait

Karen Georgia A. Thompson
21:13
30 October 2020
Olmsted Township, OH

6

Heaven-Human Harmony in Chinese Philosophy and Theology of Impurity in the Hebrew Bible

WEI HUANG

In January and February 2020, the number of Covid-19 cases in Wuhan and the surrounding provinces of Hubei, China rose quickly to a very high level. Thus, the Covid pandemic started all over the world.

As early as 23 January, the Chinese government included Traditional Chinese Medicine (TCM) treatments in the third version of the National Diagnosis and Treatment Guidelines for Covid-19.[1] Later on 31 January, the Chinese Academy of Sciences, Shanghai Institute of Materia Medica, and the Wuhan Institute of Virology issued a joint media item which claimed that a traditional Chinese medicine, *Shuanghuanglian* [双黄连] could 'inhibit' Covid-19.[2] This claim caused a stir in the public on its scientific validation since it is commonly known that TCM is not 'evidence-based medicine'. The next day, 1 February, the Shanghai Institute of Materia Medica made an official statement on its website in order to address the doubts and questioning.[3] In the statement, the institute verifies that the result is based on scientific experiments and further studies are needed to confirm the next step.

At the same time, scientists all over the world were racing to develop a vaccine for the virus. But when biomedicine had no cure against Covid-19, Chinese medical practitioners engaged a rich archive of formulae. In the new versions of the National Diagnosis and Treatment Guidelines for Covid-19 that followed, TCM treatments and formulae are continually included. (The English translations of the TCM formulae for Covid-19 may be read online; see Sun and Hsu 2020). I have no doubts that the study of Chinese medicine as science belongs to the scientists. For example, in the case of Professor Tu Youyou who won a Nobel Prize, she worked on the chemical compounds extracted from a natural plant *Qinghao* [青蒿], i.e. Artemisinin against malaria.

However, despite the controversy over how people accept TCM in fighting against Covid-19, it is not a simple scientific question. I have no intention of discussing whether TCM is better than Western/modern medicine or vice versa. I agree with the twentieth-century TCM reformer Zhang Xichun who pointed out that Chinese philosophy is the basis of TCM (Zhang 1990, p. 296). I argue that in order to understand the role of TCM during the Covid pandemic, it is necessary to analyse TCM from the perspective of Chinese philosophy.

Also, I propose to bring the Hebrew Bible into the discussion. By a cross-cultural comparison between Chinese and biblical Hebrew traditions, we might understand the reason for the acceptance of TCM during the Covid pandemic. As a consequence, this essay will show the interculturality between Chinese and Hebrew cultures.

TCM and Chinese Philosophy

Both TCM and Chinese philosophy have very complicated developments in the long history of China (Hsu 2018; Fung 1948). It is beyond the scope of this short essay to survey the history of the two. Rather, in this part, I focus on explaining several shared theoretical concepts of TCM and Chinese philosophy.

For TCM practitioners, *Yellow Emperor's Inner Canon* (*Huangdi neijing*) [黄帝内经] is the most revered medical text. It has a very complicated textual history (Harper 1998). The general edition available for us was not by a single person but was reworked and updated in its transmissions for later generations. Now it still holds the status as being a reliable source whose overall philosophy guides medical practice.

The key concepts in *Huangdi neijing* included *qi* [气] and *yin-yang* [阴阳]. These ideas appeared in very early Chinese texts. Both *qi* and *yin-yang* had already been related to health and longevity. Zhuangzi [庄子], a Chinese philosopher in the late fourth century BCE, identified *qi* as the basis of human life: 'The life is due to the collecting of the breath (*qi*). When that is collected, there is life; when it is dispersed, there is death' (*Zhuangzi*, Outer Chapters, 'Knowledge Rambling in the North', translated by James Legge). The terms *yin-yang* were also known in Chinese ancient books. Experts in *yin-yang* were contemporaries of Confucian scholars before the Qin dynasty (221–206 BCE). In the third-century text *Dao De Jing* [道德经] by Laozi, in chapter 42 it is written: 'The ten thousand creatures carry *yin* and embrace *yang*, pouring their *qi* together, thus becoming harmonious' (Linnell 2015). *Yin-yang* were the inseparable abstract aspects of all things, and the *qi* of the two (i.e. *yin* and *yang*) mix together to achieve the harmonious-ideal state of all things. This verse represents the cosmogony of Laozi (Chen 2003, p. 235). 'The ten thousand creatures' refer to all things on earth. In theory, the harmonious-ideal state of the universe is achieved by the balance of *yin-yang*.

qi permeates the universe, and it constitutes matter and causes growth and change. *qi* often takes on the qualities of the places in which it is. To take human body as an example, *qi* in the liver becomes liver *qi* and when it moves to the heart, it becomes heart *qi* (Hsu 1999, pp. 78–83); *qi* is in constant flux and flow based on the interactions of *yin-yang*. *Huangdi neijing* took these concepts and analogized the body and the universe in a system of 'correlative cosmology' (Graham 1986). The relations and analogies between the body (including the emotions) and Heaven in terms of *yin- yang* could be read in *Huangdi neijing*:

Heaven is round, earth is square; people's heads are round, their feet are square and thereby correspond to them. Heaven has the sun and moon; people have two eyes. Earth has nine regions; people have nine orifices. Heaven has wind and rain; people have joy and anger. Heaven has thunder and lightning; people have the notes and sounds. Heaven has four seasons; people have four limbs. Heaven has five tones; people have the five depots. Heaven has six pitches; people have six palaces. Heaven has winter and summer; people have cold and hot [ailments]. Heaven has ten days; people have the hands' ten fingers ... Heaven has *yin* and *yang*; people have man and wife. The year has 365 days; the body has 360 joints. ('Huang Di neijing: Ling Shu Jing* [灵枢经] 71.3; English translation by Lisa Raphals 2020)

The analogies between Heaven and body were the basis for understanding the medical theory of TCM. TCM understands the cause of the disease from its philosophical perspective. The harmonious-ideal state of the universe-body is defined as healthy. Breathing exercises, meditation, diet, herbs and other treatments are expected to adjust the *qi* in one's body, which helps harmonize *yin* and *yang*. TCM offered and emphasized control over physiological and mental processes of the body and mind as a way of self-cultivation through the transformation of *qi*. 'Self-cultivation' brings some 'moral' character in TCM. The 'moralization' of health culminated in the *Huangdi neijing*. Today, we believe that if the cause of the disease is unknown then there will be no adequate treatment.

The system of 'correlative cosmology' also applies to the society or the state. There are clear political connotations in *Huang Di neijing*, according to which the heart equals 'the official of the ruler' (*Su Wen* [素问] 8). The ruling organ in the human body parallels the ruling position in the state. There was consequently a correspondence between political philosophy and medicine. When those responsible for the well-being of the state were confronted with disorder, it was essential to make appropriate policies to prevent it. When those responsible for the health of society and nation were confronted with disease, it was necessary to understand the nature of diseases and apply appropriate treatments (Unschuld 2003, p. 200). In this regard,

TCM serves much more than a professional healing method as in modern times. On the other hand, TCM functions as a system of social norm.

With the above understanding of Chinese philosophy and TCM, we move to another ancient society, ancient Israel, to explore their traditional way of understanding disease and health.

Theology of Impurity in the Hebrew Bible

In the minds of the ancient people of Israel, like many ancient people in West Asia and elsewhere, the main causes of physical and mental disorder involved spiritual forces like ghosts (Scurlock 2006) and impersonal powers such as curses, bad omens, and divine punishments. Needless to say, ethical and cultic trespasses could unleash the wrath of deities and provoke painful symptoms. It is not surprising that disease and medical treatment were generally related to some kind of magic power because the ancients lacked the scientific knowledge, pharmaceutical products, technology and so on that we have today. Different gods were in charge of epidemic diseases and medical treatment. They provided patients with diagnosis and treatment of different types of diseases (Geller 2010), and their respective temples were healing places (Scurlock 2014). There are some hints in the Hebrew Bible that suggest the existence of shamanistic healers in pre-exilic Israel (Gerstenberger 2018). Considering that the writing and formation of the Hebrew Bible has gone through a long history of redaction and transmission, its authors and editors would surely have been aware of other medical traditions in Israel's surrounding cultures.

Thus, illness was defined as the result of a violation of YHWH's established rules and also a means of punishment by YHWH. The cure of the illness means restoring the existing world order. YHWH was conceptualized as the ultimate source for disease and injury, and also the only healer for human beings in the Hebrew Bible.

Compared with their neighbours, the writers of the Hebrew Bible have a strong monotheistic tendency. Other alternative powers for healing were regarded as illegitimate. For instance, Ahaziah king of Israel got sick, then he sent his people to the temple of Baal Zebub at Ekron. Elijah cursed him for abandoning YHWH. In the end, Ahaziah died according to the word of Elijah (2 Kings 1). The biblical texts intentionally make an obvious contrast between YHWH and the 'illegitimate' sources for healing (Hos. 5.13; 2 Chron. 16.12). Only YHWH is the ultimate source for curing and healing. This world view also explains why professional physicians are almost absent in the Hebrew Bible. These human physicians might associate with other non-YHWH powers, and they were likely to gain some significant role in the society that might challenge the authority of YHWH. Thus, in the world constructed by the Yehudite literati, physicians and other sources for healing were marginalized and ignored (Ben Zvi 2018). All this said, the Hebrew Bible drew a distinction between the polytheistic health care system and Israel's own monotheistic one. Thus, the biblical text condemned divination, witchcraft and communication with the dead (Deut. 18.10–12).

As with most pre-modern medical traditions, no causal relation between microbial pathogen and disease was established in the Hebrew Bible. However, similar to the understanding of TCM, the state of being healthy in the Hebrew Bible involved treating the person using a holistic approach. Illness was not limited to the physical body. For example, in the book of Job in the wisdom literature, Job, who revered God, lost his wealth, his children and his health. All this comes from God's plan, which cannot be predicted or understood. As a patient, one just needed to believe that God is just. This is another supplementary explanation for disease in the biblical world (Avalos 2018, p. 42). Furthermore, 'healing' could also be conceptualized to the land (2 Chron. 7.14) or water (2 Kings 2.21–22; Ezek. 47.8–9). As in TCM, the 'moral' character of illness and healing is also a key feature of the Hebrew Bible.

The theology of impurity served as the theoretical basis for public health in Israelite society. In this regard, the book of

Leviticus can be viewed as an extensive manual of treatments. Leviticus has a clear conceptual standard for what causes impurity to the Israelite community. Impurity becomes the threat to right relationship. In these priesthood texts, rituals are of great importance in maintaining Israel's right relationship with YHWH.

J. Klawans identified two types of impurity according to the priestly perspective (Klawans 2000). 'Ritual impurities' include menstruation, giving birth to babies, and burying the dead improperly. All these convey an impermanent contagion with a number of natural processes. The sacrificial system serves as a ritual of purification. On the other hand, 'moral impurities' consist of incest, murder and idolatry, which are considered to be sinful and cannot be simply 'washed away' with sacrifices. As E. Gerstenberger pointed out, some passages from the Psalms might function as prayers necessarily used in certain ceremonial actions for healing (Gerstenberger 2018). In any case, both the two kinds of impurity offend YHWH.

The knowledge of medicine and healing in the Hebrew Bible was related to its basic religious framework. The theology of impurity worked as a system of social boundaries. The most common 'leprosy' in the Hebrew Bible could probably encompass a wide variety of illnesses. The priestly writer prescribed isolation in order to keep the disease from spreading. Such chronic patients were strictly restricted from entering the sacred space and excluded from the public community life. If the disease was cured, one could return to the community only after a proper sacrifice was carried out by the priest (Lev. 13–14).

Conclusion

It was not my intention to apply the theology of impurity directly to the covid situation. No patient today should be labelled as 'impure'. Neither did I intend to evaluate TCM as better than the medical lore in the Hebrew Bible, or vice versa. Generally speaking, people nowadays believe in modern medicine. We see traditional medicine as outdated and useless in the context of modern medicine. But, facing this new coronavirus with its many unknowns, we are in the same situation as people in pre-modern times. We do not (yet) know the whole story.

Biblical writers were not interested in any scientific study of disease; nor did they intend to give full records about treatments or cures of disease. In Israelite society, the theology of impurity urged society members to maintain a good relationship with YHWH through a proper sacrificial system and practices. In general, the cause of disease was believed to be the breakdown of the relationship between YHWH and the Israelites. In the Chinese ancients' minds, illness meant the imbalance between *yin* and *yang* inside human bodies, which corresponded to the inharmonious relationship between Heaven and human beings. If I may, I would be so bold as to suggest that biblical writers would understand the Chinese philosophy of TCM in this regard.

For the present moment when a new disease broke out with no effective and reliable treatment, people would go back to old traditions seeking for insights, wisdom or comfort. As what happened in China, although TCM does not belong to the scientific medicine, the Chinese authority determined to include TCM in the National Guidelines for Covid-19. To some extent, it indicated a spontaneous response from Chinese traditions to this new challenge of Covid-19. In the eyes of scientific medicine, the TCM formulae are surely unreliable because of the lack of related experimental data. Be that as it may, what TCM offered during the pandemic met the needs of people. It is the embodiment of common values in the society. Such was the case in the Hebrew Bible.

Notes

1 See www.nhc.gov.cn/xcs/zhengcwj/202001/f492c9153ea9437bb58 7ce2ffcbee1a.shtml (accessed 18.12.20).

2 See www.cas.cn/yw/202001/t20200131_4733137.shtml (accessed 18.12.20).

3 See www.simm.cas.cn/xwzx/zhxw/202002/t20200201_5496163. html (accessed 18.12.20).

References

Hector Avalos, 2018, 'Health Care in the Levant', *Journal of the Social History of Medicine and Health* 医疗社会史研究 3.2: 39–45.

Ehud Ben Zvi, 2018, 'Ancient Medicine and World Construction among the Literati of Late Persian Period/Early Hellenistic Judah', *Journal of the Social History of Medicine and Health* 医疗社会史研究 3.2: 46–57.

Guying Chen 陈鼓应, 2003, *Laozi zhu yi ji ping jie* 老子注译及评介 (A Commentary on Lao Zi), rpt. Beijing: Zhonghua shuju.

Yu-lan Fung, 1948, *A Short History of Chinese Philosophy*, New York: Macmillan.

M. J. Geller, 2010, *Ancient Babylonian Medicine: Theory and Practice*, Chichester: Wiley-Blackwell.

Erhard S. Gerstenberger, 2018, 'Notes on Healing in the Old Testament' 萨满主义:《旧约》中的医疗, *Journal of the Social History of Medicine and Health* 医疗社会史研究 3.2: 94–110.

A. C. Graham, 1986, *Yin-Yang and the Nature of Correlative Thinking*, Singapore: Institute of East Asian Philosophies.

D. Harper, 1998, *Early Chinese Medical Literature*, London: Kegan Paul International.

Elisabeth Hsu, 1999, *The Transmission of Chinese Medicine*, Cambridge: Cambridge University Press.

———, 2018, 'Traditional Chinese Medicine: Its Philosophy, History and Practice', in Hilary Callan (ed.), *The International Encyclopedia of Anthropology*, Hoboken, NJ: John Wiley & Sons.

Jonathan Klawans, 2000, *Impurity and Sin in Ancient Judaism*, Oxford: Oxford University Press.

Bruce R. Linnell, 2015, *Dao De Jing by Lao Zi: A Minimalist Translation* (accessed 9.1.21: www.gutenberg.org/files/49965/49965-pdf. pdf).

Lisa Raphals, 2020, 'Chinese Philosophy and Chinese Medicine', in Edward N. Zalta (ed.), *The Stanford Encyclopedia of Philosophy* (accessed 18.12.20: https://plato.stanford.edu/archives/win2020/entries/ chinese-phil-medicine/).

JoAnn Scurlock, 2006, *Magico-Medical Means of Treating Ghost-Induced Illnesses in Ancient Mesoptamia*, Leiden: Brill.

———, 2014, *Sourcebook for Ancient Mesopotamian Medicine*, Atlanta: SBL.

X. Sun and E. Hsu, 2020, Translation of Beijing's recommendations for Traditional Chinese Medicine (TCM) treatment of Covid-19, *Responding to an unfolding pandemic: Asian Medicines and Covid-19*, Sienna Craig, Barbara Gerke and Jan van der Valk (eds) *Special Issue. Fieldsights: Hot Spots* (accessed 18.12.20: https://culanth.org/fieldsights/translation-of-beijings-recomendations).

P. U. Unschuld, 2003, *Huang Di neijing Su Wen: Nature, Knowledge, Imagery in an Ancient Chinese Medical Text*, Berkeley: University of California Press.

Zhang Xichun 張錫純, 1990, '*Lun zhexue yu yixue zhi guanxi*' 論哲學與醫學之關係 (Concerning The Relation of Philosophy and Medicine) in *Yi xue zhong zhong can xi lu* 醫學衷中參西錄 (Records of Heart-felt Experiences in Medicine with Reference to the West), rpt. Taiyuan: Shanxi kexue jizhu chubanshe, pp. 296–8.

Zhuangzi, Outer Chapters, 'Knowledge Rambling in the North', translated by James Legge (1815–1897) (accessed 18.12.20: https://ctext.org/zhuangzi/knowledge-rambling-in-the-north).

7

Reopening the Churches and/as Reopening the Economy: Covid's Uncovering of the Contours of 'Church Theology'

GERALD O. WEST

Covid-19 has revealed much about the systemic realities of our world, both at the global and at the local levels. This essay probes what the Covid pandemic has in its brief time among us revealed about the contours of what the South African *Kairos Document* called 'church theology'.

The essay begins with a reflection on the revelatory capacity of viruses with respect to social systems, including theological systems. The essay then takes a specifically economic focus, reflecting on how Covid has exposed the economic inequality of our globalized world, and how Covid has also exposed the kind of economic analysis that undergirds 'church theology'. This leads to some reflection on the *Kairos Document*'s understanding of 'church theology', as well as subsequent post-*Kairos Document* reflections on this theological trajectory, including the related analysis of Walter Brueggemann's biblical theological trajectories and the Ujamaa Centre's kindred analysis, derived from South African Black theology, of *the Bible as a site of struggle*. These preliminary reflections lead into some detailed analysis of how Covid's presence has exposed the contours of 'church theology'. The essay concludes by reiterating a reformulated version of the *Kairos Document*'s 'challenge to the churches', summoning South African churches to turn away from (repent of) 'church theology' towards imagined forms of 'prophetic theology' that might engage the South African state on a post-Covid (belatedly post-apartheid-settler-colonial) re-envisaged redemptive and/as redistributive inclusive de-colonial economic order.

The Revelatory Capacity of Viruses

Understanding how Covid is 'doing theology' is as much a necessary work in progress as are the biomedical engagements with this new virus. Alongside Covid, the HI-virus continues to have an agentive presence among us, having 'taught, or retaught, us many important lessons', as Peter Piot says in the opening sentence of his Foreword to the *Encyclopedia of AIDS* (Piot 2001, p. xxi). What of Covid? While Covid has not been among us long enough and is not present in our bodies long enough (for those who survive), it has certainly revealed much about our global and local contexts. Like HIV before it, Covid has exposed, among many other systems, the economic systems that make our world so unequal. Covid, like HIV, has also revealed the contours of our theological systems (see also Beverley Haddad's essay, Chapter 9, in this volume). This essay argues for the recognition that economic systems and theological systems are systemically aligned. I also argue that Covid offers an opportunity never to return to the oppressive order of the economically unequal 'normal'. As Arundhati Roy aptly puts it, 'Historically, pandemics have forced humans to break with the past and imagine their world anew. This one is no different. It is a portal, a gateway between one world and the next' (Roy 2020). However, I argue, the pervasiveness of 'church theology', as the preferred theology of the post-apartheid churches and the post-apartheid state, constrains whether and how we might walk through this Covid-enabled portal.

Unequal Economic Systems

Though Covid is a global pandemic in ways in which the HIV pandemic no longer is, having been relegated to being an 'indecent' (Althaus-Reid 2000) disease (afflicting homosexuals, intravenous drug users, sex workers and Africans), Covid has uncovered global economic inequalities along with related regional and local economic inequalities (Ataguba 2020; Baldwin and Weder di Mauro 2020). From the outset, for example, in South Africa both the primary medium and the primary content of communication about Covid have been oriented to the middle classes. English-language television channels have been the preferred medium of national announcements about the government's 'State of Disaster' and its accompanying 'Lockdown' levels (with a 'trickle-down' into other local languages). Moreover, Covid content with respect to each of the core preventative protocols (washing hands, social-distancing and masks) as well as the specific restrictions of various lockdown levels has been slow to recognize and adapt to our systemically unequal but enduring apartheid-era social geography and our persisting post-apartheid reality of systemic economic inequality (Arndt *et al.* 2020; Bhorat *et al.* 2020; Swinnen and McDermott 2020).

Though it is too early to expect careful theological reflection on the economic dimensions of Covid – which is why this book project is so important – the theology that is being done indicates that the 'church theology' trajectory is pervasive. For example, in a one-page reflection on 'How Christian Theology Helps Us Make Sense of the Pandemic', even Barney Pityana, who we might have expected, given his prophetic theology history (Pityana and Villa-Vicencio 1995), to have offered a more robustly 'prophetic theology' analysis, frames his analysis in 'church theology' terms (Pityana 2020).

As I have argued extensively (West 2016, pp. 445–542), 'church theology' has become the pervasive norm, even in South Africa, a context we might have imagined would forever be marked by liberatory-prophetic forms of theology. 'Church theology' has become the preferred theological trajectory of the churches (even the previously prophetic churches) as well as the publicly approved theology of the South African post-apartheid state.

'Church Theology'

'Church theology' as an analytical concept is perhaps the South African *Kairos Document*'s most significant contribution to Christian theology. Though the *Kairos Document*'s understanding of 'prophetic theology' has received the larger share of attention over the years since 1985 (Speckman and Kaufmann 2001; Vellem 2010), it is its discernment of distinctive and contending theological trajectories within the churches that is its most profound insight. The 'church' itself was recognized as a site of ideo-theological struggle and 'church theology' was its default theological trajectory. What is often forgotten is that what has become known as *The Kairos Document*, a key theological process and product of South African 'Contextual Theology', was originally titled *Challenge to the Church: A Theological Comment on the Political Crisis in South Africa: The Kairos Document* (Kairos 1985). It is the Revised Second Edition that uses the tertiary title as the main title (Kairos 1986). *The Kairos Document* (or the *Kairos Document*), as its original 25 September 1985 Preface asserts, 'is a critique of the current theological models of the country. It is an attempt to develop, out of this perplexing situation, an alternative biblical and theological model that will in turn lead to forms of activity that will make a real difference to the future of our country' (Kairos 1985, p. i).

The theological argument of the *Kairos Document* is that 'the church' and 'theology' must be recognized as sites of struggle. 'Theology' was a site of contestation. Identified as overlapping with theological trajectories within the church but identifiably distinct was 'State Theology' (Kairos 1985, pp. 3–7). However, precisely because 'state theology' is largely located outside the church, albeit having co-opted 'its own prophets ... from the ranks of those who profess to be ministers of God's Word in some of our Churches' (Kairos 1985, p. 7), the real concern for the *Kairos Document* is the church itself. The church is not able to resist 'state theology' because it is itself captured by 'Church Theology' (Kairos 1985, pp. 8–14), an a-contextual form of Christian theology that 'instead of engaging in an in-depth analysis of the signs of our times, ... relies upon a few stock ideas derived from Christian tradition and then uncritically and repeatedly applies them to our situation' (Kairos 1985, p. 8). The 'challenge to the church', therefore, is to turn away from 'church theology' '[t]owards a Prophetic Theology' (Kairos 1985, p. 15).

Before the *Kairos Document* offers its understanding of the theological contours of 'Prophetic Theology' (Kairos 1985, pp. 15–25), it analyses 'the fundamental problem' of 'church theology' (Kairos 1985, p. 13). As I have indicated, my interest in this essay is 'church theology', for it is this theological trajectory that has become the dominant theological trajectory in post-apartheid South Africa, which Covid has confirmed. The *Kairos Document* did not anticipate this, so its focus is on 'prophetic theology', which receives most of the revision in the Second Revised Edition (Kairos 1986, pp. 17–27, 34–5 n.15). Our Covid-kairos, I argue, requires that we return to reflect on 'church theology'.

First, according to the *Kairos Document* 'church theology' 'has not developed a social analysis that would enable it to understand the mechanics of injustice and oppression' (Kairos 1985, p. 13). 'Church theology' lacks the capacity to provide a systemic-structural social analysis. Second, 'church theology' also lacks an appropriate theological strategy (besides 'neutrality') to engage the state with an appropriate political theology (Kairos 1985, pp. 13–14). The point that the *Kairos Document* is making here is that the claim of 'church theology' to political neutrality 'enables the status quo of oppression (and therefore violence) to continue. It is a way of giving tacit support to the oppressor' (Kairos 1985, p. 13). By refusing to 'take sides' with the oppressed, 'church theology' is taking sides with the oppressor (Kairos 1985, p. 13); by refusing to do theology 'from below' with the oppressed Black masses, 'church theology' is doing theology from 'the top', with the White ruling elite (Kairos 1985, p. 11). Third, more fundamentally, 'church theology' lacks these theological capacities precisely because it is grounded in 'an other-worldly' faith and spirituality that is 'purely private and individualistic' (Kairos 1985, p. 14). The *Kairos Document* reiterates this third, foundational, theological orientation in a number of different ways. For example, in its discussion of the concept of 'justice', the *Kairos Document* argues that the problem of 'structural injustice' cannot be addressed through an emphasis on 'individual conversions' for it 'is not merely a problem of personal guilt' (Kairos 1985, p. 10). Personal morality is not the problem; systemic racial and/as economic and political oppression is.

The *Kairos Document* then goes on to err in its analysis, for it concludes the chapter on 'church theology' by arguing that, 'It hardly needs saying that this kind of faith and this type of spirituality has no biblical foundation' (Kairos 1985, p. 14). The problem, apparent to biblical scholarship, is that 'church theology' is indeed a biblical theology! Having identified 'the church', 'biblical interpretation' and 'theology' as 'sites of struggle', the *Kairos Document* fails to recognize that the Bible too is a site of struggle. The Revised Second Edition of the *Kairos Document* compounds the problem by adding the following summons in its discussion of 'prophetic theology': 'Our KAIROS impels us *to return to the Bible* and to search the Word of God for a message that is relevant to what we are experiencing in South Africa today' (Kairos 1986, p. 17, original emphasis). I have reflected extensively on the *Kairos Document*'s failure to identify the Bible itself – internally, intrinsically, inherently and indelibly – as a site of struggle, drawing substantially on the work of the biblical hermeneutics of Contextual Theology's sister theological tradition, South African 'Black Theology' (West 2000; West 2017a; West 2019; West 2020). Though my primary interest concerns how the Bible is used within 'church theology', my focus in this essay is on the contours of 'church theology' itself, particularly because 'church theology' is now the preferred theological trajectory of both the South African churches and the South African post-apartheid state. Ironically, having been partially constituted by forms of 'prophetic theology', the post-apartheid state, constituted by a prophetic alliance between the African National Congress (ANC), the South African Communist Party (SACP), and the Congress of South African Trade Unions (COSATU), now prefers not to have 'prophetic theology' challenge it, choosing 'church theology' as its dialogue partner (West 2016, pp. 445–542).

Both church and state derive their 'church theology' from the Bible. 'Church theology' is a biblical theology. In order to delve more deeply into the contours of 'church theology', including indications of why the South African state would prefer this kind of theological trajectory, I turn in the next section to the early work of Walter Brueggemann, who offers us a suggestive analysis of the Bible's 'church theology' trajectory.

Two Trajectories

Walter Brueggemann proposes that we recognize two major theological trajectories that cut trans- and inter-textually through the Bible (Brueggemann 1993, pp. 202–6), even though they change their form as they find expression in different socio-historical contexts. The analytical insight of Brueggemann's proposal is that what characterizes each trajectory is its connection to particular socio-historical sites of struggle.

The two trajectories Brueggeman discerns are the 'Mosaic liberation' trajectory and the 'royal consolidation' trajectory (Brueggemann 1993; Brueggemann 1979). These two contending trajectories can be traced, which Brueggemann does, through each of the socio-historical periods of ancient 'Israel' and into the New Testament historical period. Brueggemann summarizes these two theological trajectories as follows. The Mosaic tradition, or what the *Kairos Document* speaks of as 'prophetic theology', 'tends to be a movement of protest which is situated among the disinherited and which articulates its theological vision in terms of a God who decisively intrudes, even against seemingly impenetrable institutions and orderings'

(Brueggemann 1993, p. 202). In dialogue with this trajectory is the Davidic tradition, or what the *Kairos Document* refers to as 'church theology' (and now 'state theology' as 'church theology'), which 'tends to be a movement of consolidation which is situated among the established and secure and which articulates its theological vision in terms of a God who faithfully abides and sustains on behalf of the present ordering' (Brueggemann 1993, p. 202).

What Brueggemann's analysis offers to our understanding of the contours of 'church theology' is an explicit identification of its economic dimension. The economic is implied in the *Kairos Document*'s analysis, particularly if one is familiar with the understanding of 'race as class' in South Africa, recognizing that apartheid was a form of 'racial capitalism' (Sebidi 1986, pp. 31–5; Terreblanche 2002, pp. 14–15). Within the community-based work of the Ujamaa Centre for Community Development and Research we insist that we must be overt about this economic dimension of 'church theology' (West and Zwane 2020).

In a later related essay, Brueggemann characterizes this 'consolidation' trajectory as a 'structure legitimation' trajectory (Brueggemann 1992, p. 16; Brueggemann 1985), offering us a fuller understanding of 'church theology'. He argues, astutely, that there are aspects of the 'structure legitimation' theological trajectory – 'church theology' – that are important for our human well-being, which is the enigma of 'church theology'. This theological trajectory, he argues, 'is an assertion of *creation theology*, the sense that the world is ordered and governed' (Brueggemann 1992, p. 16). 'This theology provides', he adds, 'an ordered sense of life that is lodged in the sovereignty of God, beyond the reach of historical circumstance. It is a way of speaking about God's nonnegotiable governance' (Brueggemann 1992, p. 22). Furthermore, this theological trajectory 'satisfies a religious yearning by an affirmation of providence. Not only does God govern, but there is an order that works through the processes of history, even if that purpose is not always visible' (Brueggemann 1992, p. 22).

But, Brueggemann insists, this theology of moral coherence

is open to exploitation and tends to serve the ruling classes in both church and society, each of which 'regularly identifies the order of creation with the current social arrangement' (Brueggemann 1992, p. 22). 'What starts as a statement about *transcendence* becomes simply *self-justification*, self-justification made characteristically by those who preside over the current order and who benefit from keeping it so' (Brueggemann 1992, pp. 16–17). As in contemporary South Africa, 'church theology' becomes a form of 'state theology'. For example, there is often a link between creation theology and royal theology, which can clearly be seen in David's impulse to build a temple (2 Sam. 7.2), Solomon's building of the first temple (1 Kings 3.1), the building of the second temple by the ruling classes that returned from exile in Babylon (Ezra 1.2), and Herod's temple, the site of Jerusalem Judahite elite control (Mark 11.27–13.2) (West 2011; Boer 2007). 'The temple', argues Brueggemann, 'is a characteristic way of legitimation, not only of God's governance and providential care but also of the particular form of power distribution with the present regime' (Brueggemann 1992, p. 17; see also Waetjen 1989, pp. 180–96).

The Bible as a Site of Struggle

Structure legitimation theology – 'church theology'– has as its central concepts consolidation and control (see for example Welch 1990), including the consolidation and control of scripture, asserting that scripture has a monovocal message, which is why, in contending with the Jerusalem city-temple system, Jesus also contends with the Sadducees' monovocal interpretation of scripture (Mark 12.24). Itumeleng Mosala warned us in the late 1980s that if we failed to engage the reality of biblical texts as sites of ideo-theological economic contestation, then a text's dominant ideo-theological economic orientation would be co-opted to serve the kindred ideo-theological economic orientations of South African elites (Mosala 1989, pp. 28, 188). Covid has confirmed this analysis.

Covid's Contribution

On 1 June 2020 the South African government formally
announced that the South African economy would 'reopen',
as the government moved the country from Lockdown Level
4 to Level 3 (Government 2020a). There had been no plan to
'reopen' churches on 1 June when President Cyril Ramaphosa
initially announced the shift on 24 May (Government 2020d),
even though he had met virtually on 19 May with an interreligious range of faith leaders (Government 2020c). Unexpectedly,
the President again addressed the nation on Tuesday 26 May
(Government 2020b). The sole purpose of the 26 May address
was to announce the 'reopening' of places of worship. The
array of reasons given by the religious sector in lobbying for
'reopening' religious places have not been made public. What
is clear is that Christian churches in particular played a decisive role, including the historically 'prophetic theology' South
African Council of Churches (SACC) (Tandwa 2020). From
the government announcement it would appear that permit-
ting 'religious counselling', in addition to religious leaders
being designated as 'essential workers' with respect to funerals, was a key factor (Government 2020a), though this in itself
did not require the 'reopening' of churches, mosques, temples,
synagogues, etc.

As Zapiro's cartoon, 'Faith in the Rules' (Figure 7.1) captures
rather well, the decision to reopen religious sites of worship
was odd. The primary form of pressure the churches put on the
government, it would seem, was economic (though giving in to
the religious sector's demands demonstrates the government's
desire to consolidate their alliance with the churches' predominant theological orientation of 'church theology').

Yet, no sooner had the President announced the reopening
of the churches and other places of worship than there was a
significant public backlash (Evans 2020), with many churches
and other faith communities refusing to reopen. Having pushed
the government to reopen churches, many church leaders
recognized how difficult and costly it would be to reopen
within the Covid protocols set by the state, namely:

Figure 7.1: Zapiro, 'Faith in the Rules' (2020). Published in Daily Maverick (28 May 2020 and used by permission of the artist).

We [the government] will also permit religious gatherings such as church services as of June 1, so long as health, hygiene and social distancing is observed. This means that we must maintain 1.5 metres between the maximum of 50 congregants, should the chosen venue be able to accommodate such. We must all be wearing masks when we attend our places of worship, and the washing of hands or sanitization should be undertaken prior to worship.

Our places of worship must be sanitized, and the screening of participants is mandatory. The issued directions elaborate on the other protocols that should be observed. (Government 2020a)

Economic factors prompted both the urgency to reopen and the difficulty of actually opening within governmental Covid protocols. In this respect 'reopening' the churches is akin to 'reopening' the economy. As with the rest of the economy, the sacred economy required 'reopening' churches, without which many church-leaders could not transact the business of being church. How does a church collect tithes and offerings without 'reopening'? It is not incidental that most of the churches that did open immediately were precisely those types of African Christianity which use church 'services' to extract money from members (Nakeli-Dhliwayo 2020), with much of it going to their 'men of God' church leaders (Gifford 2004, pp. 61–70). Middle-class-based churches too, however, have not ignored Covid's threat to their financial maintenance, making use of online platforms to facilitate tithing/giving (Caboz 2020).

Has the 'church theology' of the 1980s morphed, with much of African Christianity (Gifford 1998, pp. 345–8), to a form of 'church theology' in contemporary South Africa in which churches have become increasingly consolidation-theology sites of ecclesial economic self-maintenance? It seems as if even historically 'prophetic theology' churches have embraced a self-sustaining ecclesial system of institutional-economic consolidation, establishing an ecclesial version of a self-sustaining economic system of consolidation-extraction and consolidation-consumption.

Puzzled by the 'reopening' of the churches while restaurants were to remain closed, Julius Malema, the leader of the Economic Freedom Fighters (EFF) political party, asked on 28 May 2020, 'what is the contribution of the church into the economy?' (Malema 2020). Malema holds, it would seem, a 'church theology' understanding of the church as a wholly internal economic system of institutional self-maintenance rather than a 'prophetic theology' understanding of the church as a potential resource for systemic social transformation. The kind of transformative economic agenda that the prophetic sectors of the church advocated in the 1980s, reflected in both South African Contextual Theology and South African Black Theology (West 2016, pp. 559–61), are not evident in South Africa's contemporary Covid-conditioned public theological realm.

The key question that haunts the Ujamaa Centre as we witness an emergent second Covid wave is how the 'reopened' church gradually returning to the 'old normal' responds to the challenge of President Cyril Ramaphosa for a post-Covid new economic order. On Day 154 of Level 2 Lockdown, on 27 August 2020, replying to questions in parliament, Ramaphosa argued: 'Clearly this Covid moment, as I say, has given us leverage, the opportunity, to usher in a new era.' This has been a recurring theme of Ramaphosa's, recognizing in the Covid pandemic's stark revelation of South African post-apartheid reality the summons to rebuild a more just and economically equal post-colonial South Africa. Having failed to radically restructure the South African economy after apartheid, the South African government envisages the post-Covid period as an opportunity for economic redress, belatedly eradicating settler-colonial inequality. Does Ramaphosa, who himself has contributed to the state's embrace of 'church theology' (West 2016, pp. 536–42), even expect a contribution from the church, except in the sphere of individual morality? Malema clearly has few expectations that the church has anything to offer to the slogan cry of the South African masses: 'economic freedom in our lifetime'. Significantly, though at the forefront of calls for a Basic Income Grant in the 1990s (Barchiesi 2007, p. 572), the South African churches have failed to add their voices to calls for the Covid-led Temporary Employee / Employer Relief Scheme (TERS) to be transformed into a form of 'basic income grant' (Black Sash 2020). Committed to ensure their own institutional financial well-being, 'church-theology' has little to offer theologically towards systemic socio-economic well-being.

In his address to the South African nation on 16 September, six months into lockdown, Ramaphosa announced the move into Lockdown Level 1, the most minimal level. Having made the announcement, he went on to encourage South Africans to embrace the popular dance song, 'Jerusalema', which had become a South African contribution to online forms of response to Covid lockdowns.

In just over a week from now, South Africans will celebrate Heritage Day under conditions that will be better in many ways from what we have experienced over the last six months. I urge everyone to use this public holiday as family time, to

reflect on the difficult journey we have all travelled, to remember those who have lost their lives, and to quietly rejoice
in the remarkable and diverse heritage of our nation. And
there can be no better celebration of our South African-ness
than joining the global phenomenon that is the Jerusalema
dance challenge. So I urge all of you to take up this challenge
on Heritage Day and show the world what we are capable of.
Just as we have acted together to defeat this virus, we must
roll up our sleeves and get to work rebuilding our economy.
(Government 2020f)

Ramaphosa, like other ANC leaders before him, invokes
'Jerusalem'. While I do not want to make too much of the
theological impulses that may have shaped the musicians who
composed and performed the song or what may well have been
a well-meaning gesture on the part of our President towards
national solidarity and celebration, Ramaphosa is not the first
ANC leader to invoke metaphors of 'Jerusalem'. On an ANC
visit to the Soviet Union in 1927, President Josiah Tshangana
Gumede was inspired by a vision for a united front of African
nationalists, communists and workers in South Africa. This
vision prompted him to use religious imagery: 'I have seen the
world to come, where it has already begun. I have been to the
new Jerusalem' (Gumede 1927). In the 2006 Nelson Mandela
Lecture, a speech full of a confusing mix of 'church theology'
and 'prophetic theology' (West 2016, pp. 482–91), then President Thabo Mbeki also alluded to a 'New Jerusalem', drawing
overtly on William Butler Yeats' poem 'The Second Coming'.
Mbeki appeals to his audience not to allow a 'monstrous beast'
to be born from South Africa's New Jerusalem. For our country not to 'fall apart' (alluding to Achebe 1958), he argues,
ironically (for Mbeki ushered in an economic era in which the
South African economy was reconnected to global capitalism
(West 2017b)), 'we must in the first instance, never allow that
the market should be the principal determinant of the nature of
our society' (Mbeki 2006). And now Ramaphosa has invoked
Jerusalema/Jerusalem. The repetitive resolutely 'church theology' lyrics of the dance tune turn the eyes of those who

dance to the music away from the realities of systemic injustice which Covid has so starkly revealed towards the heavenly Jerusalem. *Jerusalema ikhaya lami* (Jerusalem is my home), we are summoned to sing, *Uhambe nami* (Walk with me) to heaven, *Zungangishiyi lana* (Do not leave me here) on earth, for *Ndawo yami ayikho lana* (My place is not here) on earth (Master KG 2020).

We may choose to dance to this tune, but we would do well not to embrace its world-denying individually focused 'church theology'. Would that Ramaphosa had identified a 'prophetic theology' song for us to dance to, a song which imagined a post-Covid, but Covid-revealed, economically transformed society. Paul Gifford rightly characterizes the variant forms of 'church theology' reflected in African churches across the continent when he says that this form of Christianity is a 'domesticated Christianity' which lacks the capacity to provide a serious challenge to the economic and political realm of the state (Gifford 2009, p. 215). This form of Christianity is focused on the individual, not the communal; on the personal, not the structural; on the eschatological (see Chapter 11 by Hadje C. Sadje), not the kin-dom of God on earth; 'it is not concerned with a renewed order or any "new Jerusalem"' (Gifford 1998, p. 339). We might dance together to 'church-theology' tunes, but the theological shape of their lyrics has nothing theologically prophetic to offer to post-Covid socio-economic transformation.

Post-Covid Post-Colonial Theological Vaccination

Non-systemic 'charity'-type responses by the faith-based sector have a place in demonstrating solidarity with and symptomatic care of the vulnerable. Individual morality and charity do have a place, but they can only avoid 'church theology' co-optation if they are embedded in a systemic understanding and analysis of inequality. The work of the Ujamaa Centre in collaboration with the South African government's Solidarity Fund in local communities is a good example of how charity can be integrated within a systemic analysis (Government 2020e).

A theological legacy of Covid will be that it has prompted us to reflect critically, again, on the contours of 'church theology'. Charting the contours of 'church theology' is critical if we are to develop a 'prophetic theology' vaccine, as we in the Ujamaa Centre and similar faith-based 'prophetic' projects continue to walk with 'church theology' shaped local churches towards prophetic theological resources that have the capacity to contribute to socio-economic systemic transformation of post-Covid post-colonial South Africa (see for example Chapter 14 by Sithembiso Zwane). We will dance to 'prophetic theology' 'struggle' songs with the organized formations of the poor and marginalized with whom we work towards the summons of Jesus to 'the kin-dom' of God *on earth* as it is in heaven' (Matt. 6.10).

References

Chinua Achebe, 1958, *Things Fall Apart*, London: Heinemann.

Marcella Althaus-Reid, 2000, *Indecent Theology: Theological Perversions in Sex, Gender and Politics*, London and New York: Routledge.

Channing Arndt, Rob Davies, Sherwin Gabriel, Laurence Harris, Konstantin Makrelov, Sherman Robinson, Stephanie Levy, Witness Simbanegavi, Dirk van Seventer and Lillian Anderson, 2020, 'Covid-19 Lockdowns, Income Distribution, and Food Security: An Analysis for South Africa', *Global Food Security* 26: 100410.

John E. Ataguba, 2020, 'Editorial: Covid-19 Pandemic, a War to Be Won: Understanding Its Economic Implications for Africa', *Applied Health Economics and Health Policy* 18: 325–8.

Richard Baldwin and Beatrice Weder di Mauro (eds), 2020, *Economics in the Time of Covid-19*, London: CEPR Press.

Franco Barchiesi, 2007, 'South African Debates on the Basic Income Grant: Wage Labour and the Post-Apartheid Social Policy', *Journal of Southern African Studies* 33.3: 561–75.

Haroon Bhorat, Tim Köhler, Morné Oosthuizen, Benjamin Stanwix, François Steenkamp and Amy Thornton, 2020, *The Economics of Covid-19 in South Africa: Early Impressions*. Working Paper 202004, University of Cape Town: Development Policy Research Unit (DPRU).

Black Sash, 2020, 'Basic Income Support for People Aged 18 to 59 Now!' *Amandla Awethu* (accessed 31.10.20: https://awethu.amandla. mobi/petitions/basic-income-support-for-aged-18-to-59-now).

Roland Boer, 2007, 'The Sacred Economy of Ancient "Israel"', *Scandinavian Journal of the Old Testament: An International Journal of Nordic Theology* 21.1: 29–48.

Walter Brueggemann, 1979, 'Trajectories in Old Testament Literature and the Sociology of Ancient Israel', *Journal of Biblical Literature* 98.2: 161–85.

———, 1985, 'A Shape for Old Testament Theology, I: Structure Legitimation', *The Catholic Biblical Quarterly* 47.1: 28–46.

———, 1992, 'A Shape for Old Testament Theology, I: Structure Legitimation', in Patrick D Miller (ed.), *Walter Brueggemann Old Testament Theology: Essays on Structure, Theme, and Text*, pp. 1–21, Minneapolis: Fortress.

———, 1993, 'Trajectories in Old Testament Literature and the Sociology of Ancient Israel', in Norman K. Gottwald and Richard A. Horsley (eds), *The Bible and Liberation: Political and Social Hermeneutics*, pp. 201–26, Maryknoll, NY: Orbis.

Jay Caboz, 2020, 'A South African App Is Helping More Than 1,000 Churches Migrate Online – and It Comes with Digital Tithing', *Business Insider South Africa* 12 September (accessed 17.3.21: www.businessinsider.co.za/a-south-africa-app-is-helping-churches-migrate-online-it-comes-with-online-tithing-and-1200-have-signed-up-edisciples-2020-9).

Jenni Evans, 2020, 'Lockdown: Mixed Reaction from Religious Leaders to Relaxed Rules for Gatherings', *News 24*, 27 May 2020 (accessed 17.3.21: www.news24.com/news24/SouthAfrica/News/lockdown-mixed-reaction-from-religious-leaders-to-relaxed-rules-for-gatherings-20200527).

Paul Gifford, 1998, *African Christianity: Its Public Role*, London: Hurst & Company.

———, 2004, *Ghana's New Christianity: Pentecostalism in a Globalizing African Economy*, Bloomington & Indianapolis: Indiana University Press.

———, 2009, *Christianity, Politics, and Public Life in Kenya*, London: Hurst & Company.

Government, 2020a, 'Minister Nkosazana Dlamini Zuma: Coronavirus Covid-19 Level 3 Lockdown Regulations' (accessed 16.9.20: www.gov.za/speeches/minister-nkosazana-dlamini-zuma-coronavirus-covid-19-level-3-lockdown-regulations-28-may).

———, 2020b, 'President Cyril Ramaphosa Leads Call for National Day of Prayer and Update on Provisions for Religious Sector, 26 May' (accessed 16.9.20: www.gov.za/speeches/president-cyril-ramaphosa-leads-call-national-day-prayer-and-update-provisions-religious).

———, 2020c, 'President Cyril Ramaphosa Participates in Virtual Meeting with Faith Leaders, 19 May' (accessed 16.9.20: www.gov.

za/speeches/president-cyril-ramaphosa-participates-virtual-meeting-faith-leaders-19-may-19-may-2020).

———, 2020d, 'President Cyril Ramaphosa: Developments in South Africa's Risk-Adjusted Strategy to Manage the Spread of Corona virus Covid-19' (accessed 6.10.20: www.gov.za/speeches/president-cyril-ramaphosa-developments-south-africa%E2%80%99s-risk-adjusted-strategy-manage-spread).

———, 2020e, 'Solidarity Fund' (accessed 10.11.20: https://solidarity-fund.co.za/).

———, 2020f, 'Statement by President Cyril Ramaphosa on Progress in the National Effort to Contain the Covid-19 Pandemic' (accessed 17.9.20: https://sacoronavirus.co.za/2020/09/16/statement-by-president-cyril-ramaphosa-on-progress-in-the-national-effort-to-contain-the-covid-19-pandemic-3/).

Josiah Tshangana Gumede, 1927, 'Speech of J. T. Gumede, President of the African National Congress, at the International Congress against Imperialism, Brussels, February 10–15 1927', ANC (accessed 17.9.20: www.sahistory.org.za/archive/speech-j-t-gumede-president-african-national-congress-international-congress-against).

Kairos, 1985, Challenge to the Church: A Theological Comment on the Political Crisis in South Africa: The Kairos Document, Braamfontein: The Kairos theologians.

———, 1986, The Kairos Document: Challenge to the Church: A Theological Comment on the Political Crisis in South Africa, Revised Second Edition ed., Braamfontein: Skotaville.

Julius Malema, 2020, 'Reopening the Church', 28 May 2020 (accessed 17.3.21: www.youtube.com/watch?v=7QGCalnUbDk).

Master KG, 2020, Jerusalema (accessed 17.3.21: https://afrikalyrics.com/master-kg-jerusalema-translation).

Thabo Mbeki, 2006, '4th Annual Nelson Mandela Lecture by President Thabo Mbeki: University of Witwatersrand, 29 July 2006' (accessed 17.9.20: www.dirco.gov.za/docs/speeches/2006/mbek0729.htm).

Itumeleng J. Mosala, 1989, Biblical Hermeneutics and Black Theology in South Africa, Grand Rapids: Eerdmans.

Kekeletso Nakeli-Dhliwayo, 2020, 'Money, Not Prayer, Is Reopening Churches', The Citizen 28 May 2020 (accessed 17.3.21: https://citizen.co.za/news/opinion/opinion-columns/2292243/money-not-prayer-is-reopening-churches/).

Peter Piot, 2001, 'Foreword: Lessons from the Global HIV/AIDS Epidemic', in Raymond A. Smith (ed.), Encyclopedia of AIDS: A Social, Political, Cultural, and Scientific Record of the HIV Epidemic, pp. xxi–xxiv, New York: Penguin.

Barney Pityana, 2020, 'More Eyes on Covid-19: Perspectives from Religion Studies – How Christian Theology Helps Us Make Sense of the Pandemic', South African Journal of Science 116.7/8: 1.

———— and Charles Villa-Vicencio (eds), 1995, *Being the Church in South Africa Today*, Johannesburg: South African Council of Churches.

Arundhati Roy, 2020, 'The Pandemic Is a Portal', *Financial Times* (accessed 29.10.20: www.ft.com/content/10d8f5e8-74eb-11ea-95fe-fcd274e920ca).

Lebamang Sebidi, 1986, 'The Dynamics of the Black Struggle and Its Implications for Black Theology', in Itumeleng J. Mosala and Buti Tlhagale (eds), *The Unquestionable Right to Be Free: Essays in Black Theology*, pp. 1–36, Johannesburg: Skotaville Publishers.

McGlory T. Speckman and Larry T. Kaufmann (eds), 2001, *Towards an Agenda for Contextual Theology: Essays in Honour of Albert Nolan*, Pietermaritzburg: Cluster Publications.

Johan Swinnen and John McDermott (eds), 2020, *Covid-19 and Global Food Security*, Washington DC: International Food Policy Research Institute (IFPRI).

Lizeka Tandwa, 2020, 'We Did Not Coerce Government into Allowing Places of Worship to Open – SACC', *News24*, 27 May 2020 (accessed 17.3.21: www.news24.com/news24/southafrica/news/we-did-not-co erce-government-into-allowing-places-of-worship-to-open-sacc-20200527).

Sampie Terreblanche, 2002, *A History of Inequality in South Africa, 1652–2002*, Pietermaritzburg: University of Natal Press.

Vuyani S. Vellem, 2010, 'Prophetic Theology in Black Theology, with Special Reference to the Kairos Document', *Hervormde Teologiese Studies* 66.1.

Herman C. Waetjen, 1989, *A Reordering of Power: A Socio-Political Reading of Mark's Gospel*, Minneapolis: Fortress.

Sharon D. Welch, 1990, *A Feminist Ethic of Risk*, Minneapolis: Fortress.

Gerald O. West, 2000, 'Kairos 2000: Moving Beyond Church Theology', *Journal of Theology for Southern Africa* 108: 55–78.

————, 2011, 'Tracking an Ancient near Eastern Economic System: The Tributary Mode of Production and the Temple-State', *Old Testament Essays* 24.2: 511–32.

————, 2016, *The Stolen Bible: From Tool of Imperialism to African Icon*, Leiden and Pietermaritzburg: Brill and Cluster Publications.

————, 2017a, 'The Co-Optation of the Bible by "Church Theology" in Post-Liberation South Africa: Returning to the Bible as a "Site of Struggle"', *Journal of Theology for Southern Africa* 157: 185–98.

————, 2017b, 'Religion Intersecting De-Nationalization and Re-Nationalization in Post-Apartheid South Africa', in C. Bochinger, J. Rüpke and E. Begemann (eds), *Dynamics of Religion Past and Present, Proceedings of the XXI World Congress of the International Association for the History of Religions, Erfurt, August 23–29, 2015*, pp. 69–83, Berlin: De Gruyter.

————, 2019, 'Scripture as a Site of Struggle: Literary and Socio-Historical Resources for Prophetic Theology in Post-Colonial, Post-Apartheid (Neo-Colonial?) South Africa', in Jione Havea (ed.), *Scripture and Resistance*, pp. 149–63, New York and London: Lexington/Fortress Academic.

————, 2020, 'Serving the Sighs of the Working Class in South Africa with Marxist Analysis of the Bible as a Site of Struggle', *Rethinking Marxism* 32.1: 41–65.

———— and Sithembiso Zwane, 2020, 'Re-Reading 1 Kings 21:1–16 between Community-Based Activism and University-Based Pedagogy', *Journal for Interdisciplinary Biblical Studies* 2.1: 179–207.

8

out of breath 2[1]

i form words
in order to share some thoughts with you
in order to share my feelings with you
in order to share our spirit with you

i form words out of breath
and you left me out of breath

i form words
and i will take a stand, upon your words
your words, give me breath
your words, give me thoughts
your words, give me feelings
your words, give me spirit

your words ... harmony, qi, yin, yang, purity, impurity,
    leprosy, balance
your words ... kairos, recovery, reopen, consolidation,
    legitimation, leverage
your words ... unstory, crisis, public, private, welfare, health,
    faith, wellbeing, change
your words leave me breathless
and breathlessness adds another color to my face
'I can't breathe,' those kinds of words, make me oblige
but obligations are easily shifted, yet, for me
your words, your breaths, are obliging
Uhambe nami / walk with me, those kinds of words, give me
    breath

i form words
in order to share some thoughts with you
in order to share my feelings with you
in order to share our spirit with you

i form words out of breath
and you have left us with words to breathe

Jione Havea
Wurundjeri land & waters
Kulin nations
30 October 2020

Note

1 In the spirit of Teresia Teaiwa, 'Out of breath' (read at a meeting in Nadave, Fiji, 2014).

PART 2

In Relation

9

Private and Public Pandemics: Theological Imperatives Summoned by HIV and Covid

BEVERLEY GAIL HADDAD

Health crises affect communities across the world to varying degrees. In fact, some pandemics such as MERS-CoV, HIV, or Ebola, have been largely ignored by parts of the global community due to the specificity of geographic location, or perceived risk. Covid has changed that! The pandemic has spread across the globe at an alarming rate affecting all geographic locations and communities. While it has, as with other health crises, affected the poor, it has in many ways had a greater impact in some of the richest nations of the world. Death rates have been significantly higher in parts of the developed world, with the USA recording to date the highest death rate (WHO 2020).

Before the Covid pandemic, my personal experience has been in the field of HIV. Living in the global epicentre of the HIV pandemic – KwaZulu-Natal, South Africa – it has shaped my academic, activist and priestly work for almost 20 years. In this time, I have carried out theological reflection through an engagement with the social sciences and with the lived realities of those I know personally who are HIV-positive. I have lived through the denialist years of President Thabo Mbeki (1999–2008) when HIV was a death sentence, to the present where South Africa has the largest national roll-out of anti-retroviral drugs in the world with nearly 5 million people on treatment in 2018 (UNAIDS 2020a).

In this essay, I extend this reflective HIV experience to the new challenges brought about by Covid and argue that the same three theological imperatives that I identified in the HIV pandemic nearly ten years ago, apply to our new crisis. The discussion will identify both the theological resonances and the differences brought about by the two pandemics.

Public Health Responses

To HIV

The first two cases of HIV in South Africa were documented in 1983 (Ras *et al.* 1983). Both cases involved gay men and, together with the rapid rate of HIV infection in the gay community in the USA, it was soon labelled as a 'gay disease' in (South) Africa. This perception changed dramatically as it became clear that HIV transmission in South Africa extended to the heterosexual community. Almost 40 years later, 7.5 million South Africans are HIV-positive with an official death rate of 72.000 (AVERT 2020). And it has been estimated that Mbeki's refusal to roll-out ARV treatment led to the death of over 330,000 South Africans (Chigwedere *et al.* 2008).

Despite HIV quickly becoming a serious public health issue, the South African Department of Health followed the public health approach taken in the USA which focused on a human rights approach. This meant that HIV counselling and testing was voluntary with an informed consent procedure required. HIV was treated differently from other sexually transmitted and lethal infectious diseases resulting in HIV exceptionalism (de Kock *et al.* 2002). As de Kock *et al.* argue, an exceptionalist approach would almost certainly not have been applied in the USA or Europe if an AIDS epidemic of African severity had existed in the general population (de Kock *et al.*, p. 69). They further argued that while the human rights approach protected those who were HIV positive from structural oppression such as access to services, it resulted in enormous social stigma and contributed towards a conspiracy of silence (de Kock *et al.*,

p. 69). As my activist work can attest, silence leads to fear of disclosure of HIV positive status which in turn leads to death. HIV became a 'private pandemic'.

To Covid

With the Covid pandemic (with scenes in the media of people literally dying in corridors of hospitals in Italy, Spain and New York), there is no doubt of the severity of the Covid pandemic in Europe and the USA. The public health response has been vastly different compared to the exceptionalist response to HIV.

The first case of Covid-19 in South Africa was identified on 5 March 2020 in a person returning from holiday in Italy (NICD 2020). The South African government reacted swiftly, and by 26 March the country was placed into one of the hardest lockdowns in the world (South African Government 2020). This bought valuable time to prepare health services and obtain the necessary testing kits and medical equipment. The pandemic has not been as severe as in Europe and the USA and as of 31 December 2020 the official number of deaths was at 28,469 (Department of Health, RSA 2020). However, the South African Medical Research Council has estimated that between 6 May and 29 December 2020, there were a further 71,778 unaccounted natural deaths, suggesting a higher death rate due to Covid-19 (SAMRC 2020). This more limited death rate is currently under investigation, with some attributing it to a younger population, a certain amount of 'herd immunity' due to prior exposure to other coronaviruses, and possibly the benefits of a large sector of the population being on ARVs (Cullinan 2020). What is clear, however, is that the Department of Health adopted a vastly different public health approach to Covid by following the advice of the World Health Organization and South African scientists in declaring Covid a public health emergency. This pandemic is public in every way, with the South African state declaring a national disaster and implementing legal regulations that its citizens are obliged to follow or face criminal prosecution, certainly in the early stages of the lockdown (South African Government 2020). The Department of Health broadcasts educative messages about the virus and other information such as progression of the pandemic regularly on television, government websites and social media. The president, Cyril Ramaphosa, has addressed the nation every few weeks on national television since the first case of Covid was identified.

Because of this public health approach to the Covid pandemic, church leaders have had to embrace the health protocols laid out by the government, encouraging members to abide by the regulations, close the doors of churches for many months and offer digital forms of 'being church'. This is very different from the response by churches to the HIV pandemic where religious leaders were key players in promoting stigma and thwarting public health messages.

Theological Imperatives Posed by the HIV and Covid Pandemics

Theologians have sought to respond to the ethical challenges that emerge from these pandemics. In the early days of the HIV pandemic, I, with colleagues at the University of KwaZulu-Natal, formed a Collaborative for HIV and AIDS, Religion and Theology (CHART) and worked tirelessly in our classes and in our writings to present a liberatory trajectory in theological reflection that promoted life, rather than death (Haddad 2019). Now too, we are all confronted with a similar challenge as we learn to live in this 'new normal', wondering what this means for our faith and theological reflection. In this section I will discuss three theological imperatives that emerge out of both pandemics, while resisting trying to deal with the larger theodicy question as to why God allows suffering (for a fuller discussion of this question in this volume, see Chapter 10 by Volker Küster).

Christological Imperative

Jesus Christ is central to the Christian faith and a variety of theologians shape their Christologies by their particular ideo-theological orientations. For feminists and womanists, the key Christological question is whether a male saviour has any relevance to women (see, e.g., Schüssler Fiorenza 1994; Isherwood 2002; Butler 2019). For black and womanist theologians, Christ's blackness is foregrounded (see, e.g., Grant 1989; Pobee 1992; Cone 2011; McKinnis 2016; Douglas 2019). For Latin American liberation theologians, the economic dimensions of Christ and his identification with the poor is central (see, e.g., Miguez Bonino 1984; Piar 1994; Palazzi 2008).

Steve Taylor (2020) discusses the ways in which art plays a vital and often controversial role in Christianity, including the understandings of Christ. In times of illness and death, individuals have often used this creative medium to express their understanding of what Christ means to them in their situation of suffering. Christological representations in artworks demonstrate the contextual realities of the artist and also indicate their ideo-theological orientation and their 'incipient theologies' (Cochrane 1999). This has been true in both the HIV and the Covid pandemics, and one example of an art piece emerging from each pandemic will be discussed in this section.

In the case of HIV where the pandemic has been 'private', a painting by Maxwell Lawton, 'Man of Sorrows: Christ with AIDS' is particularly noteworthy.[1] Lawton was a student in the theology and arts program at Wesley Theological Seminary, Washington DC, when he was diagnosed with late-stage AIDS in 1992 (Lawton 2020). The AIDS pandemic in the US was largely limited to gay men at the time and this community faced enormous discrimination because of their sexual orientation and the growing death rates (in this community) due to AIDS-related illnesses. Lawton created this piece in 1993 as a class project, reflecting his own experience of having AIDS and the meaning of Christ amid his suffering (Kapikian 2000, p. 18). In the painting, Christ has lesions indicating AIDS and he is naked, despairing, and wears a crown of thorns amplifying his suffering. A year after completing this class project, Lawton while an artist in residence in South Africa, completed a second version of the painting with the words of Matthew 25 etched into the background in a range of different languages, 'when you offer care to the least of these, my brethren, you are doing it to me' (Kapikian 2000, p. 18). In the second piece, Lawton powerfully indicates a Christ who has AIDS and reminds those who view the painting that Christ stands alongside those who are marginalized and discriminated against because of their illness, and expects his disciples to do the same. Lawton's Christ demonstrates unconditional love that does not stigmatize or discriminate, and experiences death as a result of AIDS. By including the words of Matthew 25 in the second version, the painting shifts the 'private' HIV pandemic into the public realm and demands that all Christians respond positively with Christ to those living with AIDS. Having said this, it is important to note that Lawton's Christ does remain alone in the painting and so continues to symbolize an individualized, 'private pandemic'.

During the Easter season of April 2020 as the Covid pandemic was peaking in Europe, a painting went viral on social media. Viewers gave the painting various titles, but no one claimed the work as their own (Taylor 2020, p. 6).[2] The image depicts an almost naked Jesus, his arm hanging limply at his side, a ventilator over his mouth and nose, medical sensors on his torso and an oxygen bottle at the side of the bed. Five medical personnel are present clothed in Personal Protective Equipment; two of them are cradling Jesus' body with one holding his hands to his head in despair suggesting that the Christ figure is dying. This artistic representation, set in the contemporary context of the Covid pandemic, evokes our immediate identification with the Christ who suffers, understands and vicariously takes on the anxiety of those of us who fear death caused by the virus. More importantly, this art piece indicates the public nature of the Covid pandemic. The dying Christ is cradled by the public health care workers who have so often been the only ones present at the death of Covid patients. Because the Covid pandemic has been so public and present in national discourse, the contextual nature of the painting resonated forcefully with the crucifixion of Christ that was being remembered during Easter when the painting went viral on social media.

Both artworks engendered outrage in the media with many naming them as 'heretical' portrayals of Christ. But for others, particularly those directly impacted by these pandemics, the two contextual depictions of Christ offer another view of God through a Christological lens that speaks directly to health pandemics infused with stigma and discrimination as well as suffering and death. Both stand over against dominant 'church theology' and offer a different view of God. (For further discussion of 'church theology' in this volume, see Chapter 7 by Gerald West.)

During the HIV pandemic, church leaders resorted to the dominant church theology of retribution (Haddad 2006; West and Zengele 2004). Culturally, retribution theology coalesces with cultural notions of illness being a sign that one has displeased the ancestors and one is ill for doing something wrong. Theologies of retribution resulted in stigma and discrimination during the HIV pandemic and church leaders became complicit in stigmatizing those who were HIV-positive, employing theological resources that led to death. Stigma was exacerbated by the private nature of the pandemic and the exceptionalist approach adopted, as discussed above. Prophetic theological challenges had to break through the silent public discourse, resulting in new theological trajectories that are still evident but remain tenuous within religious life.

The Covid pandemic has similarly summoned forth theological challenges that are emerging daily. But these, unlike in the private HIV pandemic, occupy a public space. The dominant narrative within the Covid pandemic is scientific rather than religio-cultural, as was the case with the HIV pandemic. Yet there are indications of stigma in the Covid pandemic, but there are no sexual connotations such as are associated with HIV. Communities are fearful of contracting Covid-19, leading to the ostracizing of individuals who contract the virus in some communities (Xolo 2020). The prophetic theological voice was and still is in the minority within the response to the HIV pandemic. It remains to be seen whether prophetic theological voices can appropriate the public space and contribute to the Covid crisis in such a way that saves lives rather than harnesses the forces of death.

Ecclesiological Imperative

The second theological imperative summoned forth by the Covid pandemic relates to ecclesiology. During the height of the HIV pandemic, the majority of churches did not use the opportunity to embrace new ways of doing theology and being church. Quite the opposite. Being HIV-positive was a taboo subject in churches, exacerbated by the individualized approach adopted by public health officials. When the subject of HIV was addressed, church leaders preached from their pulpits that 'AIDS is a punishment from God' (PACSA 2004). This message stigmatized those who were HIV positive, as well as rendering the pandemic even more 'private'. To disclose your HIV positive status publicly could lead to death (Gugu Dlamini Foundation 2020). This 'private' HIV pandemic with its associated sexual connotations meant that many persons living with HIV stopped going to church and found support elsewhere such as in Bible study groups with others who were HIV-positive (West 2003). Church leaders sometimes did preach a different message from their pulpit, often times evoking outrage. This was the case when in 2010 Xola Skosana, a pastor in Khayelitsha, preached that 'Jesus was HIV positive' and followed his sermon by taking an HIV test publicly in front of his congregation, leading to some of his congregants doing the same (Allie 2010). This and other prophetic voices that offer a different way of 'being church' are seldom heard in the ongoing HIV crisis.

As indicated earlier, because of the overt public health discourse and government regulations curtailing public worship in the Covid pandemic, church leaders had to quickly find other ways of 'being church'. Creative digital forms of worship, pastoral interactions, daily reflections and prayers, have mushroomed across the globe. This enabled communities to attend church services, Bible study groups and prayer meetings from the confines of their homes. Membership of any church across the globe is now possible.

The greater challenge is whether church leaders can theologize differently in Covid times in ways that bring about structural change. In June 2020, Chris Ahrends, an Anglican parish priest, wrote a piece on social media titled, 'Whither the Christian Church? In search of the "New Normal" (without the Pointy Hats)' (Ahrends 2020). Ahrends challenges the institutional church to use this opportunity to reject destructive church theology that is patriarchal and hierarchical and find new life-giving approaches to 'being church'. While I have yet to conduct sustained research into the content of Covid sermons during the height of the first wave of the Covid pandemic in South Africa between April and September 2020, I explored the ideo-theological approaches to the pandemic taken by church leaders in two Anglican congregations. I tracked the weekly sermons and their pastoral approaches, assessing anecdotally the extent to which Covid was embraced as a contextual reality.

The first congregation is a relatively affluent middle-class congregation that made weekly services, including sermons, available on YouTube and daily reflections on WhatsApp. Seldom, if ever, did these evangelical church leaders make direct reference to Covid. In addition, monthly news bulletins sent by email only addressed the pandemic as it pertained to government regulations on the closure and opening of churches, or financial matters. No mention was made of people in the congregation who were ill with Covid and there was little engagement with the enormous social and economic challenges facing surrounding marginalized communities due to the national lockdown. Instead, the focus was on traditional evangelical biblical exegesis with few attempts to apply the insights contextually.

The second is a working-class congregation led by a rector who stands in the Anglo-Catholic tradition. Here too, weekly services including sermons were available on YouTube and

WhatsApp. Sermons were contextual, often beginning with a direct reference to the Covid context and then applying biblical exegesis directly to the meaning of faith as it applied to the local situation. In addition, the rector produced a separate weekly 'notice bulletin' in which the ongoing pastoral and outreach life of the congregation was encouraged. Those who were Covid positive were named, including details of isolation, hospitalization and deaths. At one time, the notice bulletin was videoed in the rector's garden as he indicated that he was displaying Covid symptoms and self-isolating. There has also been overt engagement with the physical needs of those in the community, particularly during the winter months. In these ways, the rector has been true to his Anglo-Catholic theological roots and embraced an embodiment of the Covid realities in the life of the church. This opened space for reflection, prayer and activist engagement by members of the congregation. The Covid stigma has been eradicated by the integration and embracing of the public discourse, enabling new ways of being church to emerge.

It is not yet clear if, and how, Covid has impacted theological reflection and whether there has been structural change as a result. What is clear, however, is that Covid forces us to rethink how we do our theology and what it means to be the institutional church.

Missiological Imperative

How we view our work in the world by responding to the missiological nature of the Christian faith depends, as with the other two imperatives discussed above, on the way we do theology and our ideo-theological orientation. Traditional church theology places emphasis on personal moral sin. As already discussed, this was deadly in the context of the HIV pandemic. This form of theology blames individuals for their HIV-positive status without any recognition of the structural dimension of the pandemic (Kopelman 2002; Louw 2006; Akitunde 2009).

In South Africa, the HIV pandemic was shaped not only by individual behaviour but also by issues of gender, race, sexual orientation and class, rendering particular groups of people more vulnerable to HIV infection (Abdool Karim and Abdool Karim 2010). Addressing these structural dimensions was critical to curbing the pandemic. Prophetic theologians soon realized that their theological analysis of this public health issue had to critically engage and intersect insights offered by the social sciences (Haddad 2011). The 'private' pandemic had to engage the 'public' discourse. This did not mean that religio-cultural determinants should be ignored, but rather they needed to be brought into dialogue with social science. It was only as HIV entered public theological discourse that liberatory and hopeful forms of engaged and contextual theological reflection could emerge (West and Zengele 2004; Ward and Leonard 2008; Richardson 2009; Chitando 2009).

Covid is no different. While religio-cultural issues are important, the structural dimension of the pandemic as it pertains to issues of gender, race, sexual orientation and class is critical for engaging in theological reflection. Gender-based violence has escalated globally since the pandemic began and particularly during national lockdowns. Phumzile Mlambo-Ngcuka, the Executive Director of UN Women, stated in April 2020 that a shadow pandemic of violence against women was growing: 'Confinement is fostering the tension and strain created by security, health, and money worries. And it is increasing isolation for women with violent partners, separating them from the people and resources that can best help them. It's a perfect storm for controlling, violent behaviour behind closed doors' (Mlambo-Ngcuka 2020).

Economics too needs to be central to our theological analysis. The downturn in the global economy has led to high rates of unemployment, destitution and vulnerability to economic exploitation. As a result, economically disadvantaged communities will continue to risk contracting Covid-19 as they seek out ways to earn money and find food for their families. (For a fuller discussion on a contextual theological response to economic inequality in this volume, see Chapter 14 by

Sithembisc Zwane). There is still a need for further research on the structural determinants of vulnerability to Covid infection by the scientific community, particularly in the areas of race and sexual orientation. However, preliminary work by UNAIDS on the repression of sexual minorities during lockdowns (UNAIDS 2020b), and the Centers for Disease Control and Prevention in the United States (CDC 2020) on the structural dimensions of vulnerability in racial and ethnic minorities, point to a future research agenda for theological reflection.

New Theological Horizons

In reflecting on the private and public nature of the HIV and Covid pandemics, it becomes apparent that these health crises offer the opportunity for new theological horizons to emerge. The fact that Covid is a global phenomenon with both rich and poor nations severely impacted is unprecedented in recent history. This destabilizing of the socio-political economic order and the overt strengthening of solidarity ties within the scientific community uniquely positions prophetic theologians to challenge individualized, private forms of church theology. The Covid pandemic, global and public in every way, enables liberatory forms of theological reflection to take the public centre stage. Now, as never before, there is a window of opportunity for theologies of global solidarity that embrace the structural dimensions of the Covid pandemic to inform and transform the Christian community of faith into places of hope and life amid the threat of marginalization and death.

Notes

1 The image is available at http://masartepordios.blogspot.com/2016/07/maxwell-lawton-1956-2006.html.
2 The image is available at: https://twitter.com/JamesMartinSJ/status/1243771985045360645.

References

Salim Abdool Karim and Quarraisha Abdool Karim (eds), 2010, *HIV/ AIDS in South Africa*, 2nd edn, Cambridge: Cambridge University Press.

Chris Ahrends, 2020, 'Whither the Christian Church? In Search of the "New Normal" (Without the Pointy Hats)', *Daily Maverick*, 22 June 2020 (accessed 17.9.20: www.dailymaverick.co.za/article/ 2020-06-22-whither-the-christian-church-in-search-of-a-new-normal-without-the-pointy-hats/).

Dorcas Akitunde, 2009, 'HIV and AIDS: God's Punishment for "Sexual Perversion"? The Nigerian Experience', in Ezra Chitando and Nontando Hadebe (eds), *Compassionate Circles: African Women Theologians Facing HIV*, pp. 115–29, Geneva: WCC Publications.

Mohammed Allie, 2010, '"Jesus Had HIV": Sermon Sparks South African Fury', BBC News, 1 November 2010 (accessed 13.3.13: www.bbc.com/news/world-africa-11575773).

AVERT, 2020, 'HIV and AIDS in South Africa' (accessed 15.9.20: www.avert.org/professionals/hiv-around-world/sub-saharan-africa/ south-africa).

Sara Butler, 2019, 'Feminist Christology: A New Iconoclasm?', *The Thomist: A Speculative Quarterly Review* 83.4: 493–519 (*Project MUSE*, doi: 10.1353/tho.2019.0042).

CDC, 2020, 'Health Equity Considerations and Racial and Ethnic Minorities' (accessed 17.9.20: www.cdc.gov/coronavirus/2019-ncov/ community/health-equity/race-ethnicity.html).

Pride Chigwedere, George Seage, Sophia Gruskin, Tun-Hou Lee *et al.*, 2008, 'Estimating the Lost Benefits of Antiretroviral Drug Use in South Africa', *Journal of Acquired Immune Deficiency Syndromes* 49: 410–15.

Ezra Chitando, 2009, *Troubled but Not Destroyed*, Geneva: WCC Publications.

James R. Cochrane, 1999, *Circles of Dignity: Community Wisdom and Theological Reflection*, Minneapolis: Fortress Press.

James Cone, 2011, *The Cross and the Lynching Tree*, Maryknoll, NY: Orbis Books.

Kerry Cullinan, 2020, 'Covid-19's "Slow Burn" – Africa's Low Death Rate Puzzles Researchers', *Health Policy Watch*, 14 September 2020 (accessed 1.12.20: https://healthpolicy-watch.news/76831-2/).

Kevin de Kock, Dorothy Mbori-Ngacha and Elizabeth Marum, 2002, 'Shadow on the Continent: Public Health and HIV/AIDS in Africa and the 21st Century', *The Lancet* 360: 67–71.

Department of Health, Republic of South Africa, 2020, 'Covid-19 Corona Virus. South African Resource Portal' (accessed 1.12.20: www.sacoronavirus.co.za/).

Kelly Brown Douglas, 2019, *The Black Christ*, 25th Anniversary Edition, Maryknoll, NY: Orbis Books.

Jacquelyn Grant, 1989, *White Women's Christ and Black Women's Jesus: Feminist Christology and Womanist Response*, Atlanta: Scholar's Press.

Gugu Dlamini Foundation, 2020, 'About Gugu Dlamini' (accessed 8.12.20: https://gugudlaminifoundation.org/about-us/about-gugu-dlamini/).

Beverley Haddad, 2006, '"We Pray but We Cannot Heal": Theological challenges posed by the HIV/AIDS crisis', *Journal of Theology for Southern Africa* 125: 80–90.

——— (ed.), 2011, *Religion and HIV and AIDS: Charting the terrain*, Pietermaritzburg: University of KwaZulu-Natal Press.

———, 2019, 'HIV and Faith: Shaping the response from rhetoric into transformative social action', in L. Juliana Claassens, Charlene van der Walt and Funlola O. Oledje (eds), *Teaching for Change: Essays on pedagogy, gender and theology in Africa*, pp. 151–68, Stellenbosch: Sun Press.

Lisa Isherwood, 2002, *Introducing Feminist Christologies*, Cleveland, Ohio: Pilgrims Press.

Catherine Kapikian, 2000, 'Christ with AIDS', *The Other Side*, May and June: 18–20.

Loretta M. Kopelman, 2002, 'If HIV/AIDS is Punishment, Who is Bad?', *Journal of Medicine and Philosophy* 27.2: 231–43.

Maxwell Lawton, 2020, 'Maxwell Lawton' (accessed 17.9.20: www.maxwellawton.com/abouttheartist.html).

Daniël J. Louw, 2006, 'The HIV Pandemic from the Perspective of a Theologia Resurrectionis: Resurrection Hope as a Pastoral Critique on the Punishment and Stigma Paradigm', *Journal of Theology for Southern Africa* 126: 100–14.

Leonard C. McKinnis, 2016, 'From Christ to Black Jesus: Black Theology's Christological Move as Operative in the Black Coptic Church', *Black Theology* 14.3: 235–51.

José Miguez Bonino (ed.), 1984, *Faces of Jesus: Latin American Christologies*, Maryknoll, NY: Orbis Books.

Phumzile Mlambo-Ngcuka, 2020, 'Violence Against Women and Girls: The shadow pandemic', UNWOMEN, 6 April 2020 (accessed 17.9.20: www.unwomen.org/en/news/stories/2020/4/statement-ed-phumzile-violence-against-women-during-pandemic).

NICD, 2020, 'First Case of Coronavirus Reported in South Africa' (accessed 1.12.20: www.nicd.ac.za/first-case-of-covid-19-coronavirus-reported-in-sa/).

PACSA, 2004, 'Churches and HIV/AIDS: Exploring how local churches are integrating HIV/AIDS in the life and ministries of the church and how those most directly affected experience these', unpublished research report, Pietermaritzburg: PACSA.

Félix Palazzi, 2008, 'Hope and Kingdom of God: Christology and eschatology in Latin American liberation theology', in Stephen J. Pope (ed.), *Hope & Solidarity: Jon Sobrino's Challenge to Christian Theology*, pp. 131–42, Maryknoll, NY: Orbis Books.

Carlos R. Piar, 1994, *Jesus and Liberation: A Critical Analysis of the Christology of Latin American Liberation Theology*, New York: Peter Lang.

John S. Pobee (ed.), 1992, *Exploring Afro-Christology*, Frankfurt: Peter Lang.

G. J. Ras, I. W. Simson, R. Anderson, O. W. Prozesky and T. Hamersma, 1983, 'Acquired Immunodeficiency Syndrome. A Report of 2 South African Cases', *South African Medical Journal* 64.4: 140–2.

Neville Richardson (ed.), 2009, *Broken Bodies and Healing Communities: The challenge of HIV and AIDS in the South African context*, Pietermaritzburg: Cluster Publications.

SAMRC, 2020, 'Report on Weekly Deaths in South Africa' (accessed 1.12.20: www.samrc.ac.za/reports/report-weekly-deaths-south-africa).

Elisabeth Schüssler Fiorenza, 1994, *Jesus: Miriam's Child, Sophia's Prophet: Critical Issues in Feminist Christology*, New York: Continuum.

South African Government, 2020, 'President Cyril Ramaphosa: Escalation of measures to combat coronavirus Covid-19 pandemic', 23 March 2020 (accessed 1.12.20: www.gov.za/speeches/president-cyril-ramaphosa-escalation-measures-combat-coronavirus-covid-19-pandemic-23-mar).

Steve Taylor, 2020, 'A Covid Christology: Art, atonement and the forming of the social body in a time of pandemic', *Stimulus: The New Zealand Journal of Christian Life and Thought* (accessed 28.12.20: https://hail.to/laidlaw-college/publication/1tI5uq8/article/7aaI8MP).

UNAIDS, 2020a, 'AIDSInfo' (accessed 15.9.20: http://aidsinfo.unaids.org/).

———, 2020b, 'UNAIDS and MPact are extremely concerned about reports that LGBTI people are being blamed and abused during the Covid-19 outbreak' (accessed 30.4.20: www.unaids.org/en/resources/presscentre/pressreleaseandstatementarchive/2020/april/20200427_lgbti-covid).

Edwina Ward and Gary Leonard (eds), 2008, *A Theology of HIV & AIDS on Africa's East Coast: A Collection of Essays by Masters Students from Four African Academic Institutions*, Uppsala: Swedish Institute of Mission Research.

Gerald West, 2003, 'Reading the Bible in the light of HIV/AIDS in South Africa', *Ecumenical Review* 55.4: 335–44.

BEVERLEY GAIL HADDAD

———— and Bongi Zengele, 2004, 'Reading Job "Positively" in the Context of HIV/AIDS in South Africa", *Concilium* 4: 112–24.

WHO, 2020, 'Coronavirus Disease (Covid-19) Dashboard' (accessed 1.12.20: http://covid19.who.int/).

Nomfundo Xolo, 2020, 'Covid-19: Concerns Mount Over Covid-19 Stigma in KZN', *Spotlight*, 21 May 2020 (accessed 17.9.20: www.spotlightnsp.co.za/2020/05/21/covid-19-concerns-mount-over-covid-19-stigma-in-kzn/).

10

Interpretation Against: What if not Punishment?

VOLKER KÜSTER

Susan Sontag (1978) has argued against interpretation of illness and the punishment metaphor based on her own experiences as a cancer patient. This essay explores how people nevertheless keep interpreting Covid-related matters with the help of cultural and religious resources and how at the same time the virus interprets us. It develops liberational theological perspectives and calls for an intersectional *pandemethics*.

The Plague

In my school days, Albert Camus' *The Plague* was a popular text in religious education classes to explain what theodicy – the question why God allows suffering – is about (Camus 1950). When the Covid pandemic gained momentum Camus' book had a renaissance and was sold-out in Germany, France and Italy. It took weeks before publishers came up with new editions to meet the demand.

Camus gives a dense description of how the plague affects the Algerian coastal city Oran. Among the often-ambiguous characters are the Jesuit priest Paneloux and the medical doctor Bernard Rieux, who have drawn the attention of theologically informed readers. There are two key scenes in particular. The first is a sermon by Paneloux in a Mass dedicated to St Rochus, the Saint of the Plague, at the end of a prayer week organized by church authorities in response to the epidemic. Paneloux starts with the exclamation, 'Calamity has come on you, my brethren, and, my brethren, you deserve it.' What follows is a classical penitential sermon.

The second scene is set at the deathbed of a child, 'a grotesque parody of crucifixion', where Rieux and Paneloux meet. Rieux snaps at Paneloux 'Ah! That child, anyhow, was innocent, and you know it as well as I do!' Paneloux confronts Rieux afterwards. Their positions seem irreconcilable. While Rieux is ready to fight illness, death and evil even in the absence of hope facing the absurdity of creation, Paneloux is full of hope in God's grace amid all hopelessness. Paneloux is concerned about salvation, Rieux about healing and saving lives. Yet Rieux tells the priest, 'What I hate is death and disease, as you well know. And whether you wish it or not, we're allies, facing them and fighting them together … God Himself can't part us now.'

Contagion

People also turned to Hollywood for guidance. Steven Soderbergh's 2011 thriller *Contagion* was sold out in electronic stores and traded at high prices on the internet.[1] While moderately successful after its release in 2011, all of a sudden it was seen as a prophetic movie. Soderbergh, an independent filmmaker later working in Hollywood, did not direct one of the usual catastrophes, devastation, mass panic and personalized suffering movies of the disaster genre but concentrated on how national and international health institutions and their doctors try to manage a pandemic. His crew did extensive research on the earlier SARS pandemic (2002–03) and had scientific advisers present throughout the shooting of the film.

The film starts with cool distant scenes showing how infected people around the globe touch surfaces and leave their viral fingerprints, before eventually breaking down. The captivating electronic soundtrack draws the viewer into the dynamics of the movie. It differs from Camus' story in that religion does not play a role in the plot, either as horizon of interpretation or through its agents. The only reference occurs when a doctor informs Mitch Emhoff (played by Matt Damon) that his wife, who is patient zero, has died; Emhoff does not want to believe it but gets upset instead. The doctor refers him to a hospital chaplain – without even mentioning the profession – 'There are grief counselors who are very helpful with this kind of passing. Ok? You might find some resolution there.'

Beth Emhoff had committed adultery with a former lover during her stopover in Chicago; he also gets infected and dies. The doctors who reconstruct the spread of the virus feel pity for Mitch Emhoff that the last thing his wife did in life was to cheat on him. This classical case for Christian moralists is an invitation to talk about God's punishment, but nothing is uttered of that kind.

There is a conspiracy theorist (played by Jude Law), who becomes an influential blogger; agents from the department of homeland security, who suspect bio-terrorism; wicked brokers and capitalists who want to speculate on the vaccine; but the protagonists of the film are the doctors engaged in managing the crisis and finding a cure.

Soderbergh plays with Hollywood's role clichés by casting an Afro-American and women as his heroes, while putting a white man – Mitch Emhoff (Matt Damon alias Jason Bourne) into the role of a deceived husband. Dr Ellis Cheaver (played by Laurence Fishburne), the head of the Centers for Disease Control and Prevention in Atlanta, sends his colleague Dr Erin Mears (played by Kate Winslet – the unforgotten *Titanic* diva) to Minneapolis, the hometown of the Emhoffs, to explore the situation. When she eventually gets infected, he tries to get her out, but the special plane is used by Homeland security to move an infected senator out of Chicago instead.

One can see Cheaver trying to take care of the people for whom he is responsible. A key storyline to portray his character is the interaction with a cleaner. When he talks to his fiancé on the phone telling her to leave Chicago quietly, he all of a sudden realizes that the cleaner has overheard the conversation. On leaving he says, 'I got people too, Dr Cheaver. We all do.' In an earlier scene when they met casually in the basement garage, he asked the doctor for advice regarding the suspected ADHD case of his son. Meanwhile Homeland security accuses Cheaver of treason because his wife told her friend who spread the news on social media. The underlying conflict is between national security and humanitarianism. At the end Cheaver appears at the shabby home of the cleaner and injects the son with his ration of the vaccine, telling the cleaner 'He'll be safe'. Only then does he drive home to vaccinate his wife.

Dr Leonora Orantes (played by Marion Cotillard) is sent to China by the WHO to trace the origins of the virus; she is kidnapped by people of a village that is affected most. When she is released in exchange for vaccines and finds out that actually placebos were delivered, she identifies with her kidnappers. Again, the conflict between in this case international politics and humanitarian concern flares up. Dr Ally Hextall (played by Jennifer Ehle), another of Cheaver's colleagues, who succeeded in isolating the fictitious virus MEV-1, finally injects herself with the potential vaccine to shortcut the procedure of testing and getting it licensed. In the film those in the US who are 'system relevant' are vaccinated while the vaccine is distributed through a lottery system by birthdate among the population. What is happening to the rest of the world remains an open question, but Dr Orantes has made her point.

The end of the film shows how the virus came into the world and how Beth Emhoff got infected. A bat, chased up by a bulldozer of the AIMM Alderson mining company she works for, drops a bit of banana over a pig stable. The cook who prepares the pig that ate it, shakes Beth Emhoff's hand after just rubbing his hands on his apron. It is clear to the viewer that this can easily happen again. This resonates with the end of Camus' *The Plague* when Dr Rieux reasons about the fact that the virus never dies and is still out there to resurface at a given moment. While Camus deals with an epidemic Soderbergh tackles a pandemic. Both consider viruses as part of nature and their protagonists try to save as many of their fellow human beings as they can. While Camus addresses the absurdity of human life, Soderbergh keeps it strictly rational. The illness is neither metaphorized nor stigmatized in either case. In a similar

pragmatic way Prof. Hendrik Streek, co-author of the Heins-
berg study, one of Germany's first Corona hotspots during the
carnival season in North Rhine-Westphalia, argues that we
will have to integrate Covid-19 into our virus repertoire and
learn how to live with it (Jacobs 2020). German sociologist
Ulrich Beck introduced the concept of world risk society as a
framework to understand globalization; terrorism, pandemics,
ecological disasters are all risks that go with it (Beck 1999).

Two things are striking: people are no longer referring to the
Bible and the Christian God but rather to a French philosopher
of the absurd and to popular culture to interpret the pandemic,
and they refer to the plague that has been more or less extinguished rather than to the late modern scourge HIV/AIDS.

HIV/AIDS revisited

Susan Sontag (1933–2004), a New York cultural critic, became
famous for her paradoxical polemics against interpretations
of illness and the HIV/AIDS pandemic. After being diagnosed
with cancer, she decided to write a book that was not one of
the usual 'how I defeated the illness' stories but a manifesto
against prevalent interpretations and the triad of othering,
metaphorization and stigmatization of illness (Sontag 1978), if
mere denial is not the chosen strategy.

Ten years later, after the outbreak of the HIV/AIDS pandemic
that affected many of her friends, she revisits her earlier book
and writes again against interpretation (Sontag 1989). In the
case of cancer, doctors will often tell the relatives rather than
the patient; while with the HIV/AIDS-positive, they will not
tell the relatives (Sontag 1989, p. 121). In both cases the illness
is often regarded as self-induced, due to psychological weakness in the case of cancer or sexual libertinism and perversion
in the case of HIV/AIDS infection. The latter is categorized as
the illness of a sexually stigmatized group of homosexual men
and of addicts who are consuming intravenous drugs (Sontag
1989, pp. 111–12). The HIV/AIDS pandemic among hetero-
sexuals in Africa (more than 70 per cent of those infected and

the highest death toll) is neglected. Obviously, this stigmatization is so imprinted in our cultural memory that people
rather refer to the Plague as a frame of reference in the current
pandemic, even though there would have been lessons to learn
from the HIV/AIDS pandemic in spite of all obvious differences between the diseases (Schock 2020).

Struggle of Interpretations

In the face of the Covid pandemic the strategy of othering
is widely applied as is that of mere denial. Trump speaks
about the 'Chinese virus', and in Africa it is the 'illness of the
whites'.[2] While at the beginning of the pandemic Asians were
stigmatized and harassed in the US and occasionally in Europe,
the illness itself is not stigmatized nor metaphorized. Everybody can catch it. Celebrities and politicians publicly speak
about their infection and the course of the disease. Church
leaders were quick to announce that Covid-19 was not God's
punishment. Even in American evangelical and fundamentalist
circles that argument was not popular, which leaves it to some
Christian extremists and the IS. American evangelicals turn to
Apocalypticism or entertain conspiracy theories instead. At
the same time a secular version of the punishment metaphor
is deployed by climate change activists like Greta Thunberg.

 If the punishment metaphor for illness is rejected right away
even though biblical texts are more ambiguous than modern
church leaders and theologians allow, we should be aware that
God is not at the disposal of human reasoning and the question remains: How do we respond theologically to Covid-19
from our societies and religious communities? Biblical texts
like the Psalms, Job, the Gethsemane story and Christ's outcry
on the cross 'Why have you forsaken me?' point to bringing
one's grief before God, even doubting and questioning God.
The vulnerability of human beings is mirrored in the presence
of God in the suffering on the cross. While Jürgen Moltmann
(1973) has interpreted this as a suffering within God, Johann
Baptist Metz emphasizes that we also suffer from God (2006,

p. 18). Evil, illness, aging, suffering and death are part of God's creation too. Unlike Camus' Pater Paneloux, Moltmann stands for a theology of hope that is grounded in a realized eschatology that invites people to engage in working for the reign of God that is already present in the here and now even if God is still coming. The anthropological turn of liberation theologies focusing on vulnerability and suffering let third-generation black theologian Anthony Pinn to argue for a Black Humanism over against the Christocentric approach of most liberation theologies (Pinn 2014; Pinn 2012). Miguel de la Torre to the contrary keeps being a theologian and social activist in the light of a theology of hopelessness.

> The hopelessness I advocate rejects quick and easy fixes that may temporarily soothe the conscious of the privileged but is no substitute for bringing about a more just social structure that is not based on the disenfranchisement of the world's abused. But this hopelessness that I advocate is not disabling; rather, it is a methodology that propels toward praxis. All too often the advocacy of hope gets in the way of listening and learning from the oppressed. To sit in the reality of Saturday is to discover that the semblance of hope becomes an obstacle when it serves as a mechanism that maintains rather than challenges the prevailing social structures. But this is never an excuse to do nothing. It may be Saturday, but that's no justification to passively wait for Sunday. The disenfranchised have no option but to continue their struggle for justice regardless of the odds against them. They continue the struggle, if not for themselves, for their progeny. (de la Torre 2018)

This is the ambiguity in which we have to respond to the Covid-19 crises. The new triad 'social distance, sanitize and wear your mask' demand that religious communities adapt their ritual practices. It certainly does not mean that the care for those in need should be neglected out of misunderstood hygiene concepts.

Toward an Intersectional *Pandemethics*

In order to understand the pandemic better intersectionality is a helpful tool. Unlike with the HIV/AIDS pandemic, young sexually active people who party in Ischgl, Palma de Mallorca ('Ballermann') or Berlin get infected but they don't die of the virus. Rather, they spread it. The old people and people with a history of chronic illness are the ones who died under the first wave of the virus in affluent Western European countries. Intergenerational solidarity is therefore crucial. Yet when it came to treatment age was a decisive factor for the triage: who gets access to one of the rare respirators and who does not? In the health care system of the US a class aspect is added, the rich have the access to better treatment. On top of that also a race factor occurs, more black people than white people are dying percentage wise from Covid-19. Because of their eating habits, junk food and soft drinks, drug abuse and the poverty-stricken crowded housing facilities they were prone to a grave course of the illness. George Floyd's imploration 'I can't breathe' sounds like the epitaph of a black person who is poverty-stricken, racially discriminated against, suffering from malnutrition, chronically ill and carrying the virus in his body. Unfortunately, black lives matter *less* in the Covid-19 pandemic.

In Latin America, right-wing governments use hygiene politics to prevent the spread of the virus to get rid of political dissidents. In India and other countries of the global South, migrating dayworkers got stranded in the lockdown and lost their income, which is life-threatening in the absence of social security systems. In the meantime, while a lockdown is not applicable under these circumstances, global solidarity asks for the distribution of masks and soap in situations where hundreds of people share one water tap and toilet; where huge families live in one room, social distancing is just not possible.

A postcolonial perspective raises the question whether the Western hygiene strategies are transferable to the global South or make a mockery of the living conditions there. In the global South, more people die of malaria and AIDS because they don't have access to medical treatment, and children die of starvation rather than of Covid-19. People in the global South thus exclaim, 'We'd rather die of Covid-19 than of starvation!' While in the West the ethical question regarding medical decisions like triage is 'How can the individual doctor decide which patient has a better life expectancy?', in the global South the ethical question is whether applying a certain Covid-19 measure like a lockdown costs more life than not: which one is the lesser evil? Intercultural ethics has to be aware that different contexts may not only raise different ethical problems but may also require different ethical solutions. At the same time certain ethical standards should be transcultural. How to negotiate this is an ethical challenge in itself.

How do Johann Baptist Metz's (2007) plea for a theodicy-sensitive God-talk and his constant appeal to remember the victims resonate in the context of the Covid pandemic? The announcement that 'Covid-19 is not God's punishment' in a certain sense is a late modern version of Bonhoeffer's 'cheap grace'. What is asked for instead is that churches humble themselves and stand in solidarity with the victims. We are in need of a liberation theology that analyses the intersectionality behind the pandemic and develops a habitus which allows to respond to the intergenerational, intercultural and interreligious challenges (Küster 2010) in dealing with the pandemic and at the same time has the courage to raise its prophetic voice and point out that Covid-19 is not the single risk that is shaking our planet.

It is not only we who are interpreting the virus; it is also the virus that is interpreting us. The disease has revealed the inequalities in our lives, the risks we are facing while the media coverage and the responses by governments around the globe threaten to cover them up again. Even if we find a vaccine, human beings will still have to die and injustice regarding the distribution of wealth and chances for human flourishing as well as global warming and its consequences for the whole of creation will all remain an issue.

Notes

1 Trailer available on YouTube (accessed 18.3.21: www.youtube.com/watch?v=4sYSyuuLk5g).
2 The politics of deception and deferral of the Chinese government cost global health prevention precious time. Astonishingly, against the usual stereotypes due to the heavy outbreaks in Europe and the US, it soon became the 'illness of the whites' who travel more and enjoy affluent lifestyles that make them careless for others (see Schaap 2020). European governments have also been hesitant in introducing counter measures, while American president Trump or Brazilian president Bolsonaro even responded with ignorance and neglect. All of these made the pandemic worse.

References

Ulrich Beck, 1999, *World Risk Society*, Cambridge: Polity Press.

Albert Camus, 1950 [French 1947], *Die Pest*, Hamburg: Rowohlt. English quotations are from the translation by Stuart Gilbert (London: Hutchinson, 1967).

Miguel A. de la Torre, 2018, 'Embracing Hoplessness' (lecture, accessed 18.3.21: www.youtube.com/watch?v=oCoRFMe_vgQ).

——, 2019, *Burying White Privilege: Resurrecting a Badass Christianity*, Grand Rapids: Eerdmans.

Philipp Jacobs, 2020, 'Zu viele Ängste wurden geschürt', *Saarbrücker Zeitung*, 5 June (accessed 18.3.21: www.saarbruecker-zeitung.de/nachrichten/politik/inland/virologe-hendrik-streeck-es-wurde-mir-zu-viel-gewarnt_aid-51517347).

Volker Küster, 2010, *Einführung in die interkulturelle Theologie*, Göttingen: Vandenhoeck & Ruprecht.

Johann Baptist Metz, 2006, *Memoria Passionis: Ein provozierendes Gedächtnis in pluralistischer Gesellschaft*, Freiburg: Herder.

——, 2007, *Faith in History and Society: Toward a Practical Fundamental Theology*, translated by J. Matthew Ashley, New York: Crossroad.

Jürgen Moltmann, 1973, *The Crucified God: The cross of Christ as the foundation and criticism of Christian theology*, Minneapolis: Fortress.

Anthony B. Pinn, 2012, *The End of God-Talk: An African American humanist theology*, New York: Oxford University Press.

——, 2014, 'Humanism in African American Theology', in Katie G. Cannon and Anthony B. Pinn (eds), *The Oxford Handbook of*

African American Theology, pp. 280–91, New York: Oxford University Press.

Fritz Schaap, 2020, 'Corona in Afrika: "Die Menschen haben Angst, sich mit der Krankheit der Weissen anzustecken"', *Der Spiegel*, 10 July (accessed 18.3.21: www.spiegel.de/consent-a-?targetUrl=https%3A%2F%2Fwww.spiegel.de%2Fpolitik%2Fausland%2Fcorona-in-afrika-die-menschen-haben-angst-sich-mit-den-krankheiten-der-weissen-anzustecken-a-ffdabfe9-873f-450e-a4a6-ed6136978d62).

Axel Schock, 2020, 'Corona und HIV: (K)Ein Virus für alle', *magazine. hiv*, 7 May (accessed 18.3.21: magazin.hiv/2020/05/07/corona-und-hiv/).

Susan Sontag, 1966, *Against Interpretation*, New York: Farrar, Straus & Giroux.

———, 1978, *Illness as Metaphor*, New York: Farrar, Straus & Giroux.

———, 1989, *Aids and Its Metaphors*, London: Penguin.

11

'Stripping the Thief in the Night': Decolonizing Pentecostal Eschatology during Covid

HADJE C. SADJE

Early Western philosophers criticized the spiritual dimension of eschatology, especially the Christian doctrine of the 'last things' (Fergusson 1997). Inspired by the Enlightenment, they maintained that the role of eschatology was diminishing and attempted to remove its significance in future ideal society, focusing instead on practical approaches and solutions throughout the Western world. However, according to Judith Wolfe, 'the Enlightenment did not eradicate the importance of eschatology as a structuring frame of historical and moral thought, but merely changed it' (Wolfe 2018, p. 55). In fact, eschatology continued to remain a subject of debate for contemporary social thinkers. For instance, a growing number of scholars, including Western philosophers and social scientists, have become interested in eschatological topics (Paipais 2018; Friedman 1986; Ahmad 2012; Rattansi 1982). As the term 'eschatology' evolved over time, it became secularized (David 2000, p. 254; Collins 2003, p. 64).

Eschatology is one of the major branches of Christian theology, but it means different things to different Christian groups. The two dominant approaches are 'future-oriented' and 'present-oriented' eschatology (Frey 2011, p. 6). For 'future-oriented' groups, eschatology is the Christian teaching of the 'final events of human history' or 'the last things'. For them, Christian eschatology covers such subjects as the second coming of Christ (*Parousia*), the judgement, the destruction of Satan's kingdom, the inauguration of the Kingdom of God, heaven and hell (Frey 2011, p. 6).

On the other hand, the 'present-oriented' group argues that eschatology should emphasize the present life (Frey 2011, p. 7). For them, an eschatological message must have a tangible and realistic content (Ratzinger 1988, p. 3). Today, contemporary scholars shifted the emphasis from opposition to the dialectic between the future and present (Metz 1994, pp. 613–20). They argue that eschatology is both 'future-oriented' and 'present-oriented' (Metz 1994, p. 616). Nevertheless, Christian eschatology remains a difficult subject among laypeople in the church.

Christian Eschatology as a Single all-Encompassing Framework

The Covid pandemic has elicited various responses from different groups and classes, including conservative Christian groups and the Christian eschatological discourse. The idea of Christian eschatology achieved prominence in political-economic, socio-cultural and ecological discourses over several decades, especially during the establishment of the modern state of Israel, the Y2K bug problem, the 9/11 attacks and the threat of global warming (Harding 1994, pp. 14–44). Despite its thorny symbols and technical intricacies, eschatology is one of the most widely (ab)used Christian teachings today (Verbeke, Verhelst and Welkenhuysen 1988, pp. 1–48). In fact, Christian eschatological thought continues to be a common response to contemporary global issues.

Most conservative Christian groups often approach and interpret Scripture using current global events. For example, prior to the outbreak of Covid-19, Nashville-based LifeWay Research published a research paper showing that a large number of pastors view signs of the end times through current events such as famines, war, conflicts, earthquakes and other natural disasters (Earls 2020). The Covid pandemic made this kind of eschatological thought popular again. It provides a single all-encompassing framework for understanding and explaining the global pandemic among conservative Evangelical and Pentecostal groups (Casper 2020; Cooper 2020; Gagné 2020). For example, Cami Oetman (2020), a popular international Bible speaker, claimed that Covid-19 is a sign of the end times by citing passages from Matthew 24. Oetman was not alone. US-based evangelists Paul Begley, Greg Laurie and John Hagee argue that the pandemic is God's punishment. Therefore, people must repent, practise self-piety and turn back from their sinful ways.

One particular group place strong emphasis on eschatology – the Oneness Pentecostal movement. Most Oneness Pentecostals predict world events based on their interpretation of biblical texts. For example, at the end of December 1999, many Oneness Pentecostals claimed that the Y2K problem was the sign of the end of the world (McConnell 2016, pp. 10–11; Hookway 1999). The Oneness Pentecostal preachers contended that the Y2K problem is the imminent fulfilment of 'end-time Bible prophecy' (McConnell 2016, p. 10; Hookway 1999). However, the Y2K problem with its futuristic expectations and apocalyptic speculations turned out to be a failed prediction (Poloma, pp. 1–33).

Like many other conservative Christian denominations, Oneness Pentecostals perceived eschatology as one of the central 'Apostolic' teachings that shape their social behaviour and public engagement. Although Oneness Pentecostal eschatology takes different forms, the most popular one is rapture theology or what is commonly known as 'a thief in the night' (1 Thess. 5.2). According to this view of eschatology, 'rapture' expresses the idea that faithful Christians, both living and dead, will 'be caught up', 'vanished' or 'snatched' from the earth (King 2002, pp. 1–10). Adherents of this eschatological view believe that faithful Christians will be sucked into the air to meet the Lord Jesus Christ (Ladd 1978, p. 84). This notion of rapture underlined the idea of the 'great escape' from great persecution (or the 'great tribulation') under the authority of the Antichrist (Combs 1998, p. 87). On the other hand, those who are left behind are sinners who will suffer, be condemned and be punished.

A variety of movies, most notably the 2014 apocalyptic film *Left Behind*, reintroduced, popularized and presented vivid pictures of the doctrine of rapture (Guest 2012). *Left Behind* originated as an apocalyptic novel series for young adults that made two American writers Tim LaHaye and Jerry Bruce Jenkins extremely popular. In 2016, *The Washington Post* claimed that LaHaye's *Left Behind* series has reached 10 million readers (Smith 2016). Considering the impact of literature in modern society, LaHaye and Bruce advanced and propagated the doctrine of rapture (Guest 2012). Although officially the *Left Behind* series presented itself as fictional accounts, the film has carried a subtle argument about the Christian doctrine of rapture. Much earlier, Sydney Watson published his trilogy of rapture novels, *In the Twinkling of an Eye* (1910), *Scarlet and Purple* (1913) and *The Mark of the Beast* (1915). Another early source is the *Scofield Reference Bible* (1909/1917) which strategically placed commentaries by Cyrus I. Scofield next to the biblical text. Scofield's commentaries framed the ideological framework of dispensationalism, particularly the theological thought of John Nelson Darby, which will be discussed in the next section.

These fictional novels and *Scofield Reference Bible* inspired a generation of Christian authors who collectively made a huge impact on the eschatological vision of Christian public engagement today. Rapture as a compelling storyline has evolved into a highly successful religious ideological framework (Guest 2012). However, it is important to note here that the rapture of the faithful believers is a separate event from Jesus' second coming.

The doctrine of rapture is part of an eschatological model known as dispensationalism (Ryrie 1965, p. 28). Although the relationship between rapture and dispensationalism was questioned, debates over the literal or distorted interpretation of Scripture by various dispensationalists remain at the heart of present-day eschatological discourse. Generally speaking, dispensationalism refers to the grand narrative that God has different ways of salvation for human beings during different periods of history. This kind of Christian eschatology is a product of a literal interpretation that provides a structure for Christian beliefs and, more importantly, gives meaning to a Christian's personal experiences of global current events. However, dispensationalism as a single and all-encompassing framework of Christian eschatology has produced a dangerous, corrupt and half-baked theology (Clark 2014).

Dispensational Theology: Its Open Secret, Contradictions and Consequences

The re-emergence of dispensationalism among conservative Evangelicals and Pentecostals indicate that global Christianity has not quite outgrown this religious ideology despite its numerous criticisms (Ice 1994, pp. 9–10). In fact, this eschatological model does not *supersede* but rather *competes* with other religious eschatological models. Many fundamentalist Christian preachers in Western and non-Western nations continue to quote this religious ideology (Ice 1994, p. 10).

To address this challenge, I divide this section into two parts. First, I address the questions: Why does dispensational theology receive so much criticism? Why is this eschatological thought considered to be a dangerous version of the Christian eschatology? I explore four implications of dispensationalism to sort out its hidden violence, which gave rise to heated debate among Christian theologians. Second, I address the question: Why is decoloniality (or decolonial intervention) relevant and crucial in responding to present-day Christian eschatological challenges?

Dangers of Dispensational Theology

We begin with subtle forms of violence embedded in the theology of dispensationalism. First, dispensational theology tends to compartmentalize the gospel by emphasizing a sharp distinction between the bodily and spiritual needs of the people. This is a tragic impact of the Western/Greek dualistic worldview on Christian thought. It implies that the Christian message can be privately engaging but publicly irrelevant. However, this violence-stoking rhetoric is an age-old tactic of Western/ European colonial rulers. History shows again and again that this future-oriented eschatological model permitted Western missionaries to favour the saving of 'individual souls' at the expense of the welfare of the larger community. This eschatological model still resonates today, especially in the political and economic challenges brought about by the Covid pandemic. Most Christian groups that embraced this future-oriented eschatology are indifferent towards social issues, like global poverty, human rights and climate change (Bloomquist 2009). Due to end-time Bible prophecy obsession, doomsday preachers undermine public health law and public health policy. In fact, most conservative Christian groups endanger their members by holding church services and defying government measures to prevent the spread of the coronavirus (Wilson 2020).

Second, dispensational eschatology fails to address unjust social structures. It fails to explain and challenge the evil which arises from unjust social structures within human society, focusing instead on individual wickedness. These preachers tend to blame people for being poor, underprivileged and marginalized (Zauzmer 2017). For example, during the outbreak of Covid, these preachers viewed the global pandemic as the result of individual fault regardless of race, class, gender and disability (Wilson 2020). Despite numerous studies which show that poor, coloured people are the most vulnerable victims of the pandemic, these doomsday preachers trivialize the reality of individual sin, but neglect sin and the evil in institutions and structures of contemporary society (Beyer 2020; Noppert 2020; Ro 2020).

Third, this eschatological model does not promote solidarity with nature. Instead, it tolerates ecological destruction. A case study in 2015 shows that Christian groups in the US and Australia are the least engaged with climate change issues, compared with their Buddhist counterparts. Furthermore, another study shows that conservative Christian groups are among those who are more sceptical about climate change issues. Although these conservative Christian groups were convinced that climate change is real, they viewed climate change or global warming as part of God's plan (Gander 2019). Well-known American pastor and prolific author John McArthur publicly stated, 'God intended it as a disposable planet' (Braterman 2020). Although not a dispensationalist theologian, McArthur and many conservative Christian groups have become complacent about ecological justice. Many Pentecostal groups, dispensationalists in particular, preferred to join the 'climate change skeptical group' (Clifton 2009, p. 121). For them, 'there is no pressing need to respond to it in light of the impending peril predestined for earth and her (ungodly) inhabitants' (Rakes 2016, p. 3; Althouse 2014, p. 76). Such religious ideology, scholars contend, normalized ecological exploitation and destruction (Donozo, Tutor and Guia 2019). It divorces human beings from nature. More importantly, it serves the main agenda of global capitalism, a system that maintains the lucrative flow of goods and resources.

Fourth, this version of eschatology produces a triumphalist attitude. In numerous cases, religion has been instrumental with a certain efficacy in legitimizing the colonization of indigenous people (Peterson 2001). Today, the religious argument plays a huge role in the public realm, including political discourse, public opinion and public policy (Blum 2018). Irvine Anderson (2005, p. 117) and Grace Halsell (1986, pp. 164–5) expose the hidden violence of dispensationalism through the military occupation and building of settlements in the Palestinian territories, the illegal detention of Palestinian minors, and the imposition of Israel's 'apartheid regime'. Anderson argued that Christian Zionism and their 'eschatological views' or 'end times beliefs' are shaping Western foreign policy in the Middle East. The use of dispensationalism as a single all-encompassing eschatological model offered theological legitimacy of white supremacy, racism, military occupation and colonialism (Masalha 2007; Pappé 2015). It justifies ethnic cleansing, the apartheid state and the Zionist colonialization project (illegal settlement and annexation) of the West Bank and Gaza strip (Nielsen, pp. 1–7; Sizer 2004, pp. 106–7; Pappé 2007, pp. 1–38). Despite overwhelming evidence of human rights violations, dispensationalists consistently justify Israel's brutal military occupation of the Palestinian territories (Shindler 2000; Haija 2006).

As a whole, this kind of eschatological thought shows the concerted expression of the Western imperialist/colonial ideology. Dispensationalism as a religious ideology normalizes, empowers and encourages the following: violence, a dualistic worldview, authoritarian leadership, white supremacy, racism, ethnic cleansing, apartheid state, domination, exploitation, colonialism and imperialism, regardless of whether this is done consciously or not (Owen, Wald and Hill 2006). With that said, it is important to challenge this popular eschatological thought.

Decoloniality Can Save Us!

In what ways can the decolonial option help Christian eschatology? Why is decoloniality important in addressing the hidden violence of dispensational theology?

Why decoloniality? Any introduction to the ideas of 'decoloniality' runs the risk of being too concerned with the direct political consequences of a Western political economy. As a result, the decolonial option is perceived as another form of postcolonialism or postmodernism. But this is not the case for decolonial thinkers. Decoloniality is a holistic evolutionary call for independent thinking by delinking from the Western corpus of knowledge as the main source of legitimate knowledge (Mignolo 2007). Western coloniality or 'colonial matrix of power' (Mignolo 2007, p. 156) refers to 'four interrelated domains: control of economy (land appropriation, exploitation of labour, control of natural resources); control of authority (institution, army); control of gender and sexuality (family, education) and control of subjectivity and knowledge (epistemology, education and formation of subjectivity)'. Decoloniality is thus the discernment and radical engagement of 'coloniality' as an epistemic foundation (main source) of Western colonialism. It is a radical epistemological break with

HADJE C. SADJE

the Western colonial matrix of power (Mignolo 2017, p. 2; Ndlovu-Gatsheni 2013, pp. 37–64).

I offer three reasons why decoloniality is important in interrogating popular Christian eschatological thought, particularly dispensationalism. First, decoloniality has the ability to name the problem. Decoloniality identifies dispensationalism as part of 'coloniality' or the Western 'colonial matrix of power'. This religious ideology that dominates Christian eschatological thought does not have a name for most American Christian groups. Naming and articulating the Western colonial matrix of power in the public conversation helps many Christian groups to understand and reclaim eschatology from imperial ideological captivity. More importantly, it becomes possible to contest and dismantle the imperial ideological captivity of Christian eschatology when it is exposed as socially constructed religious categories by a particular group, rather than expressions of natural facts.

Decolonial thinkers contend that it is important to indicate the difference between Western modernity/colonialism and coloniality (Mignolo 2017, p. 2). Accordingly, 'colonialism is an unhappy situation advancing modernity vision and ideals whereas, "coloniality" is the darker side of Western modernity' (Mignolo 2017, p. 2). These two concepts, 'Western modernity/colonialism' and 'coloniality', are not separated but are 'two sides of a single coin' (Grosfoguel 2010, p. 28). In the context of Christian eschatological thought, Western 'coloniality' or 'colonial matrix of power' is deeply embedded in the DNA of dispensationalism. Although dispensationalists reject modernity due to its erosion of a valid belief in the idea of certainty, I argue that dispensationalism is essentially linked to Western 'coloniality' or 'colonial matrix of power'.

Coloniality is the epistemological foundation of dispensationalism. The unquestioned and unchallenged Western epistemological foundation of dispensationalism produce a religious world view that normalizes, empowers and encourages the following: a world that contains an undesirable state of affairs, a Eurocentric morality, a predisposition to a dualistic worldview, patriarchal thinking, the religion of white supremacy, the dichotomy between body and soul, the use of violence as a last resort, exploitation, and the view of colonialism and imperialism as God's plan (De La Torre 2019, pp. 115–48). Following Mignolo's notion of decoloniality, the hidden violence of dispensationalism continues to dominate, evolve and shape the Christian eschatological thought unless 'coloniality' is properly identified and addressed. Thus, Christian eschatological thought must reconstitute its epistemological foundation outside the prevailing eschatological models of Western European and American categories and epistemologies.

Second, decoloniality as epistemic disobedience or delinking from the Western canon of knowledge eventually creates a space that values a non-Western world view, primarily the indigenous world views (Mignolo 2017, p. 9). By engaging the sources of knowledge (epistemology), decoloniality is concerned with how and where knowledge is produced. It moves away from polarities – for example, body/soul, religious/secular, communist/capitalist, traditional/modern – and thus views reality as a whole, integral, and intersectional. Hence, decoloniality sees the intersectionality of knowledge production, a religious world view, human sexuality, ecological issues and political-economic problems (Mignolo and Tlostanova 2006; Grosfoguel 2010). Decoloniality questions the geopolitics of knowledge production, with important implications for pursuing epistemic justice. Epistemic justice exposes misrepresentation of a knowledge system or world view that has been internalized as universal and absolute truth. For example, Western scholars viewed indigenous world views as pagan, primitive, backward and disposable, but Western knowledge as scientific, logical, rational and universal (Ndofirepi and Gwaravanda 2019). However, decolonial thinkers contend that indigenous world views and epistemologies are holistic, integral, relational and complex, which affirms that human beings are fundamentally part of the ecological system (McGuire-Adams 2020, pp. 34–47; Gélinas and Bouchard 2014). More significantly, decoloniality offers an understanding of the interrelationship between the self and nature, which is basically an understanding of our own self (Escobar 2006).

Third, decoloniality refuses the depoliticization and misframing of social concerns (Mignolo 2002). Like other decolonial thinkers, Mignolo contends that decoloniality is both a political and epistemic project (Mignolo 2017, pp. 360–87). Critics argue that depoliticization of social concerns in Western and non-Western countries today is worrisome (Grant 2016). For instance, for many Western academics, theological and religious discourses have no political and economic connections. Consequently, Christian theology and religious studies have no other social impacts, for example, in the policymaking process, peacebuilding, political-economic issues and public health issues – these issues are thus omitted from the Western colonial matrix of power discourse. Hence, global poverty, neoliberal ideology, white supremacy, racism, colonialism and imperialism remain largely unquestioned.

Conclusions

Christian eschatological thought is not exempt from the logic of Western imperial ideological captivity. As pointed out above, the 'Western colonial matrix of power' is embedded in the popular Christian eschatological thought of dispensationalism. A decolonial interrogation of the political-social consequences of popular Christian eschatology throws light on hidden violence of dispensationalism. Due to the weakness of biblical-theological and sociological imaginations, most Christian groups have succumbed to a literalist interpretation of eschatology. This is evident with episodes of mass hysteria and apocalyptic anxiety during and after the outbreak of Covid. Instead of taking a rational approach, most conservative Christian groups imagined the end of the world. 'The basic thought is that it is easier to imagine the end of the world than the end of capitalism' (Žižek 2012, p. 1; Jameson 1994, p. xii).

Mass hysteria and apocalyptic anxiety are used as weapons by dispensationalist preachers to prevent people from engaging in public issues. These preachers, actually, understand the power of emotion and apocalyptic imagination. While such conservative Christian groups are often portrayed as marginal extremists, the eschatological thought they espouse has influenced mainstream Western politics (particularly the geopolitical issue in the Middle East) far more than most critics realize. Decolonizing Christian eschatology is therefore crucial.

What we need is a new framework that is independent of the Western colonial matrix of power, especially in the post-Covid-19 era. It requires paradigmatic break or epistemic disobedience from the categories of Western modern and imperial languages (classification and spatio-temporal hierarchies), in which the categories of the non-Western are rearticulated. Hence, this process of epistemic reconstitution requires Christians to find new concepts, develop decolonial languages, abandon the imperial fantasy and embrace intercultural diversity and religious pluralism (Grosfoguel 2010). Hopefully, decolonizing Christian Pentecostal eschatology will create a more attentive, sensitive and responsive approach to the global socio-economic-political crisis and, more importantly, will provide a path to reconstruct post-Covid society.

References

Aijaz Ahmad, 2012, 'Three "Returns" to Marx: Derrida, Zizek, Badiou', *Social Scientist* 40.7/8: 43–59.
Peter Althouse, 2014, 'Pentecostal Eco-Transformation: Possibilities for a Pentecostal Ecotheology in Light of Moltmann's Green Theology' in A. J. Swoboda (ed.), *Blood Cries Out: Pentecostals, Ecology, and the Groans of Creation*, pp. 116–33, Eugene OR: Pickwick Publications.
Irvine Anderson, 2005, *Biblical Interpretation and Middle East Policy*, University Press of Florida.
Don Beyer, 2020, *The Impact of Coronavirus on the Working Poor and People of Color*, US Senate Joint Economic Committee (accessed 19.3.21: www.jec.senate.gov/public/_cache/files/bbaf9c9f-1a8c-45b3-816c-1415a2c1ffee/coronavirus-race-and-class-jec-final.pdf).
Karen L. Bloomquist, 2009, *God, Creation and Climate Change: Spiritual and Ethical Perspectives*, Geneva: Lutheran University Press.
Jason N. Blum, 2018, 'Public discourse and the myth of religious speech', *Journal of Contemporary Religion* 33.1: 1–16 (DOI: 10.1080/13537903.2018.1408259).

Paul Braterman, 2020, '"God intended it as a disposable planet"': meet the US pastor preaching climate change denial', *The Conversation* (accessed 19.3.21: https://theconversation.com/god-intended-it-as-a-disposable-planet-meet-the-us-pastor-preaching-climate-change-denial-147712).

Jayson Casper, 2020, 'Middle East Christians Grapple with Apocalyptic Pandemic: Covid-19 offers eschatology experts opportunity to refine public understanding of what Revelation teaches', *Christianity Today* (accessed 19.3.21: www.christianitytoday.com/news/2020/june/coronavirus-end-times-middle-east-jets-pandemic-eschatology.html).

Fred Clark, 2014, 'Bad Theology makes for bad movies: "Left Behind" is a story with no place for real humans', *Patheos* (accessed 19.3.21: www.patheos.com/blogs/slacktivist/2014/10/03/bad-theology-makes-for-bad-movies-left-behind-is-a-story-with-no-place-for-real-humans/).

Shane Clifton, 2009, 'Preaching the "Full Gospel" in the Context of Global Environmental Crises', in Amos Yong (ed.), *The Spirit Renews the Face of the Earth: Pentecostal forays in science and theology of creation*, pp. 117–34, Eugene, OR: Pickwick Publications.

John J. Collins, 2003, 'From Prophecy to Apocalypticism: The Expectation of the End', in Bernard J. McGinn, John J. Collins, and Stephen J. Stein (eds), *The Continuum History of Apocalypticism*, pp. 65–88, New York: Continuum International Publishing.

William W. Combs, 1998, 'Is Apostasia in 2 Thessalonians 2:3 a Reference to the Rapture?', *Detroit Baptist Seminary Journal* 3: 63–87.

Michael Cooper, 2020, 'What Covid-19 Has to Do With The "End Times"', *LifeWay: Facts and Trends* (accessed 19.3.21: https://factsandtrends.net/2020/04/23/what-covid-19-has-to-do-with-the-end-times/).

Stephen T. David, 2000. 'Eschatology', in *Concise Routledge Encyclopedia of Philosophy*, London: Routledge.

Miguel A. De La Torre, 2019, *Burying White Privilege: Resurrecting a Badass Christianity*, Grand Rapids: Eerdmans.

Arnold B. Donozo, Julius B. Tutor and Kim S. Guia, 2019, 'A Church's Response to the Earth's Healing in the Face of Ecological Crisis', *Bedan Research Journal* 4: 191–214 (accessed 19.3.21: www.researchgate.net/publication/337335719_A_Church%27s_Response_to_the_Earth%27s_Healing_in_the_Face_of_Ecological_Crisis).

Aaron Earls, 2020, 'Vast Majority of Pastors See Signs of End Times in Current Events', *LifeWay Research*, 7 April (accessed 19.3.21: https://lifewayresearch.com/2020/04/07/vast-majority-of-pastors-see-signs-of-end-times-in-current-events/).

Arturo Escobar, 2006, 'Political Ecology of Globality and Differences', *Gestión y ambiente*, 9.3 (Medellín): 29–44.

David Fergusson, 1997, 'Eschatology' in *The Cambridge Companion to Christian Doctrine* C. Gunton (ed.), pp. 226–44, Cambridge: Cambridge University Press.

Jörg Frey, 2011, 'New Testament Eschatology – an Introduction: Classical Issues, Disputed Themes, and Current Perspectives', in Jan G. van der Watt (ed.), *Eschatology of the New Testament and Some Related Documents*, pp. 1–32, Tübingen, Germany: Mohr Siebeck.

George Friedman, 1986, 'Eschatology vs. Aesthetics: The Marxist Critique of Weberian Rationality', *Sociological Theory* 4. 2: 186–93 (doi:10.2307/201887).

André Gagné, 2020, 'Coronavirus: Trump and religious right rely on faith, not science', *The Conversation* (accessed 19.3.21: https://theconversation.com/coronavirus-trump-and-religious-right-rely-on-faith-not-science-134508).

Kashmira Gander, 2019, 'What evangelical Christians really think about climate change', *Grist: Holy Land Studies* 5.1: 75–95.

Claude Gélinas and Yves Bouchard, 2014, 'An Epistemological Framework for Indigenous Knowledge', *Revista de Humanidades de Valparaíso Año* 2.2: 47–62.

Carl A. Grant, 2016, 'Depoliticization of the language of social justice, multiculturalism, and multicultural education', *Multicultural Education Review* 8.1: 1–13 (doi: 10.1080/2005615X.2015.1133175).

Ramón Grosfoguel, 2007, 'The epistemic decolonial turn', *Cultural Studies* 21.2: 211–23 (doi: 10.1080/09502380601162514).

———, 2010, 'Decolonizing Post-Colonial Studies and Paradigms of Political-Economy: Transmodernity, Decolonial Thinking, and Global Coloniality', *Transmodernity: Journal of Peripheral Cultural Production of the Luso-Hispanic World* 1.1: 17–48.

Mathew Guest, 2012, 'Keeping the End in Mind: Left behind, the Apocalypse and the evangelical', *Literature and Theology* 26.4: 474–88.

Rammy M. Haija, 2006, 'The Armageddon Lobby: Dispensationalist Christian Zionism and the Shaping of US Policy Towards Israel-Palestine', *Holy Land Studies* 5.1: 75–95 (doi: 10.1353/hls.2006.0006).

Grace Halsell, 1986, *Prophecy and Politics*, Westport, CN: Lawrence and Hill Company.

Susan Harding, 1994, 'Imagining the Last Days: The Politics of Apocalyptic Language', *Bulletin of the American Academy of Arts and Sciences* 48.3: 14–44.

James Hookway, 1999, 'Y2K-Bug Fears Still Plague Philippines, Despite Efforts', *The Wall Street Journal* (accessed 19.3.21: www.wsj.com/articles/SB944161653337049578).

Thomas Ice, 1994, 'Dispensational Hermeneutics', in *Issues in Dispensationalism*, Wesley Willis and John Master (eds), Chicago: Moody Press.

Fredric Jameson, 1994, *The Seeds of Time*, New York: Columbia University Press.

Paul L. King, 2002, 'Premillennialism and the Early Church', in K. Neil Foster and David E. Fessenden (eds), *Essays on Premillennialism*, pp. 1–10, Camp Hill, PA: Christian Publications.

George Eldon Ladd, 1978, *The Last Things: An Eschatology for Laymen*, Grand Rapids: Wm. B. Eerdmans.

Nur Masalha, 2007, *The Bible and Zionism: Invented Traditions, Archaeology and Post-Colonialism in Palestine-Israel*, London: Zed Books.

Walter McConnell, 2016, 'CIM/OMF and the Pentecostal/Charismatic Movement', *Mission Round Table* 11.2: 4–14.

Tricia D. McGuire-Adams, 2020, 'Paradigm shifting: centering Indigenous research methodologies, an Anishinaabe perspective', *Qualitative Research in Sport, Exercise and Health* 12.1: 34–47 (doi: 10.1080/2159676X.2019.1662474).

Johann Baptist Metz, 1994, 'Suffering unto God', *Critical Inquiry* 20.4: 611–24.

Walter Mignolo, 2002, 'The Geopolitics of Knowledge and the Colonial Difference', *The South Atlantic Quarterly* 101.1: 57–96.

———, 2007, 'Introduction', *Cultural Studies* 21.2–3: 155–67 (doi: 10.1080/09502380601162498).

———, 2017, *The Darker Side of Western Modernity. Global Futures, Decolonial Options*, Durham, NC: Duke University Press.

——— and Madina Vladimirovna Tlostanova, 2006, 'Theorizing from the Borders Shifting to Geo and Body-Politics of Knowledge', *European Journal of Social Theory* 9.2: 205–21.

Sabelo J. Ndlovu-Gatsheni, 2013, *Coloniality of Power in Postcolonial Africa: Myths of Decolonization*, Senegal: Codesria.

Amasa Philip Ndofirepi and Ephraim Taurai Gwaravanda, 2019, 'Epistemic (in)justice in African universities: a perspective of the politics of knowledge', *Educational Review*, 71.5, 581–594 (doi: 10.1080/00131911.2018.1459477).

Craig Michael Nielsen, 2012, *Israel-Palestine: A Christian Response to the Conflict*, Amsterdam: Foundation University Press.

Grace A. Noppert, 2020, 'Covid-19 is hitting black and poor communities the hardest, underscoring fault lines in access and care for those on margins', *The Conversation* (accessed 19.3.21: https://theconversation.com/covid-19-is-hitting-black-and-poor-communities-the-hardest-underscoring-fault-lines-in-access-and-care-for-those-on-margins-135615).

Cami Oetman, 2020, 'The Second Coming of Jesus: The Bible Reveals Covid-19 is a Sign', *Unlocking Bible Prophecy: Adventist World Radio* (accessed 1.10.20: https://bible.awr.org/intro/).

Dennis E. Owen, Kenneth D. Wald and Samuel S. Hill, 1991, 'Authoritarian or Authority-Minded? The Cognitive Commitments of Fundamentalists and the Christian Right', *Religion and American Culture: A Journal of Interpretation* 1.1 (Winter), pp. 73–100 (doi: 10.1353/hls.2006.0006).

Vassilios Paipais, 2018, '"Already/Not Yet": St Paul's Eschatology and the Modern Critique of Historicism', *Philosophy and Social Criticism* 44.9: 1015–1038.

Ilan Pappé, 2007, *The Ethnic Cleansing of Palestine*, Oxford: Oneworld Publications.

——, 2015, *Israel and South Africa: The Many Faces of Apartheid*, London: Zed Books Ltd.

Anna L. Peterson, 2001, 'Review: Indigenous Culture and Religion before and since the Conquest', *Latina American Research Review* 36.2, 237–54 (accessed 16.8.17: http://www.jstor.org/stable/26920 98).

Margaret M. Poloma, 2000, 'The Spirit Bade Me Go: Pentecostalism and Global Religion', paper presented at the Association for the Sociology of Religion Annual Meetings, 11–13 August, Washington, DC (accessed 19.3.21: hirr.hartsem.edu/research/pentecostalism_poloma art1.html).

Daniel Rakes, 2016, 'Pentecostalism and the Environmental Crisis: Is a Theological Climate Change Necessary?' Regent University School of Divinity (February 23), pp. 1–15.

Ali Rattansi, 1982, *Marx and the Division of Labour: Contemporary Social Theory*, London: Macmillan.

Joseph Ratzinger, 1988, *Escatology, Death and Eternal Life*, translated by Michael Waldstein, Washington, DC: The Catholic University of America.

Christine Ro, 2020, 'Coronavirus: Why some racial groups are more vulnerable', BBC Future (accessed 19.3.21: www.bbc.com/future/article/20200420-coronavirus-why-some-racial-groups-are-more-vulnerable).

Charles Ryrie, 1965, *Dispensationalism Today*, Chicago: Moody Press.

Colin Shindler, 2000, 'Likud and the Christian Dispensationalists: A Symbiotic Relationship', *Israel Studies* 5.1, The Americanization of Israel, Spring: Indiana University Press, pp. 153–82.

Stephen Sizer, 2004, *Christian Zionism: Road-map to Armageddon*, Nottingham: Inter-Varsity Press.

Harrison Smith, 2016, 'Tim LaHaye, evangelical author of 'Left Behind' bock series, dies at 90', *The Washington Post* (accessed 19.3.21: www.washingtonpost.com/entertainment/books/tim-lahaye-evangelical-author-of-left-behind-book-series-dies-at-90/2016/07/25/1f20 d3a4-5286-11e6-b7de-dfe509430c39_story.html).

Werner Verbeke, Daniel Verhelst and Andries Welkenhuysen, 1988, *The Use and Abuse of Eschatology in the Middle Ages*, Leuven: KU University Press.

Jason Wilson, 2020, 'The rightwing Christian preachers in deep denial over Covid-19's danger', *The Guardian* (accessed 19.3.21: www.theguardian.com/us-news/2020/apr/04/america-rightwing-christian-preachers-virus-hoax).

Judith Wolfe, 2018, 'The Eschatological Turn in German Philosophy', *Modern Theology* 35.1: 55–70.

Julie Zauzmer, 2017, 'Christians are more than twice as likely to blame a person's poverty on lack of effort', *The Washington Post* (accessed 19.3.21: www.washingtonpost.com/news/acts-of-faith/wp/2017/08/03/christians-are-more-than-twice-as-likely-to-blame-a-persons-poverty-on-lack-of-effort/).

Slavoj Žižek, 2012, *Year of Dreaming Dangerously*, London: Verso.

12

out of touch

Do not touch, for on that day you will surely die
the fruit hanging from the centred
tree will give you knowledge
but God forbids ... do not touch or you
will pay the consequences of knowledge
knowledge is power, power is dangerous
and danger ... don't touch

Fast forward
do not touch or you might contract those
other pandemics which many straight folk spread
do not touch reality, but look up to the illusions
with which Christian thieves steal land
wealth, bodies, faith and imagination
do not touch, or you will face punishment and isolation

Slowly, go back to the garden, that place
that story, thought to be the genesis of punishment
go back, and ...
yep, they were not alone
they had company, who helped them see things
they talked, she listened
she learned and was delighted
with possibilities and alternatives
she picked the fruit, she touched and
she ate and her eyes were opened
she saw the pleasures of knowledge
she shared – and he too took according
to his own will, he too ate
but he did not learn enough, yet, still

They touched, ate and learned
that they were naked, exposed
vulnerable, limited, disposable

In touching they irritated, and made the G*d mad
kick them out, was the G*d's solution
install the cherubim, flash the
flaming sword, lock the garden
guard the other fruit
so that no mortal may live forever
that other fruit that gives life eternal
keep that one, out of touch

Should i say amen to that?

Living is limited, no mortal lives forever
but life is generational, life passes
from one generation to the next
living will end, but life passes through
diseases will infect, pain and stop the living
but life is not out of touch, and life
should not be out of touch, for generations to come
life should not be out of touch, for all creatures and plants
in the sea, under the ground, above the skies and on earth
life should not be out of touch
to that, i say AMEN

Jione Havea
Wurundjeri land & waters
Kulin nations
31 October 2020

13

Vulnerability: Embodied Resistance During Covid

KUZIPA NALWAMBA

Love your hands! Love them. Raise them up and kiss them. Touch others with them, pat them together, stroke them on your face ... Love your mouth ... This is flesh ... Flesh that needs to be loved. Feet that need to rest and to dance; backs that need support; shoulders that need arms, strong arms ... Love your neck; put a hand on it, grace it, stroke it and hold it up. And all your inside parts that they'd just as soon slop for hogs, you got to love them. The dark, dark liver – love it, love it, and the beat and beating heart, love that too. More than eyes or feet. More than lungs that have yet to draw free air. More than your life-holding womb and your life-giving private parts ... love your heart. For this is the prize. (Morrison 1987, p. 88)

The quote above is from Toni Morrison's Nobel prize-winning novel, *Beloved*. In it, the protagonist, Baby Suggs, once ministered to fugitives and former slaves to whom she gave a message to transcend the Christian message of self-renunciation in this life and deliverance after death. She 'encouraged former slaves ... to linger over the free black body – a body so easily reviled, broken, discarded, assaulted and commodified while enslaved – and to love it as flesh' (Smith 2012, p. 71). Her sermons sought to re-establish the spirit–body relationship as a locus for 'restoration' – a paradoxical act of resistance that, at once, comes to terms with the traumatic experience of enslavement, naming its wounds and in the same token drawing strength from it.

A Paradox

A vulnerability hermeneutic must develop alongside the naming, social critique of, and action against structural violence and oppression in society. It must take into account the paradoxical nature of vulnerability as both threatening and enriching. As such, a reflection on the liberating potential of vulnerability cannot be removed from our concrete lived reality. Recent studies in vulnerability have rightly highlighted the need for interdisciplinary reflection on the subject in order to complexify and provide 'thickness' for theological engagement with the subject.

The Covid pandemic as a threat to human life has heightened awareness of our shared vulnerability. The trauma of disease and death is real. Even so, Covid has also revealed the power of life in human creativity, resilience and solidarity. Such multidimensional, real and even paradoxical experiences of human life help us to reflect more sharply on our experience. Such reflections give us pause to do theology without rushing to a neat conclusion or declaration that oversimplifies real concrete human experiences of vulnerability.

Although vulnerability is universal and a condition of being human, vulnerability is experienced in varied ways. The experience of poor people who lost their jobs at the onset of the Covid pandemic and became exposed to hunger and homelessness overnight cannot be compared to that of the wealthy who checked out of their regular lives without any looming threat to their well-being. We therefore make a distinction between vulnerability as a human condition and vulnerability as 'situated or contextual' (Springhart 2017, p. 17) experience premised on one's social location. The Covid experience has exposed the situated vulnerability that is on a continuum with vulnerability as a human condition (i.e. that all life is susceptible to atrophy).

Elizabeth O'Donnell Gandolfo describes vulnerability as the condition that makes possible the connection with 'the power of love, beauty ... (and) the redemptive power of divine love itself' (O'Donnell Gandolfo 2015, p. 3). She sidesteps categories that attempt to explain why God permits evil in the world, preferring to outline the human condition as laden with 'unavoidable possibility of harm' (p. 38). Marilyn McCord Adams' acerbic observation that 'God has created us radically vulnerable to horrors, by creating us as embodied persons, personal animals, enmattered spirits in a material world of real or apparent scarcity such as this' (Adams 2006, p. 37) is instructive.

The 'anthropological constants' (Schillebeeckx 1980, p. 733) of embodiment, relational dependence, ambiguity and mortality entail that we live constantly on the verge of harm, as a human condition. In that regard, vulnerability is the matrix within which we negotiate life and also, more importantly, experience God's redemption (O'Donnell Gandolfo 2015, p. 34). Being dependably tethered to our vulnerability, we long and search for healing, a life without pain, suffering and fragility. That deep-seated desire to escape our own vulnerability may be the starting point for the exploitation of other people's and creation's vulnerability. And yet if we consider vulnerability a hermeneutical tool, it becomes a theological resource that unlocks our perspectives on the work of God's spirit; the incarnation, cross and resurrection of Jesus Christ; God's relationship with creation and the meaning of the church as the (vulnerable) body of Christ.

Bodies and Structures

I was diagnosed with presumed Covid-19 on 18 March 2020, two days after Switzerland effected its first partial lockdown. At that time, testing was not available to everyone. If you had symptoms you called a hotline, described your symptoms and answered a few questions to establish your profile. If the case was deemed not serious based on the cluster of symptoms, age range and pre-existing medical condition, standard advice for self-care was pronounced with caution to get treatment if the condition changed. Although my condition was diagnosed as not to be serious, it turned out to be a drawn-out case – what we termed 'long Covid'. I spent March to the end of July struggling with illness every single day. After two months, I phoned the Covid-19 call centre again to check if I qualified for a test, given how long the symptoms had persisted. I still did not meet the criteria.

My struggle with bodily pain and disorientation made me feel trapped in my body. Despite that, I got up each day, exercised and did whatever amount of work I could manage. Although I cannot attribute my recovery to that routine, it was certainly a strategy for bodily resistance at a time of great vulnerability. When an illness overtakes the body, one is keenly reminded, as I was, that we are embodied and that the body is an aspect of our being, even if it is not the whole of our being. In the truest sense of understanding that quest for freedom, that body is the locus of liberation.

My existential bodily limitation, awareness of pain and consciousness that I was feeling that pain inspired this reflection. As a sentient, conscious being, bound up in relationships of vulnerability and interdependence, I reflected not only on my own pain but also the pain of others and how my frailty was affecting others. Being far from home, I worried about how possible death would affect my family, especially my elderly parents.

The fact that I and many others who got infected in the early months of the pandemic could not get the care we hoped for exposed one kind of vulnerability. I was in my first year of living in a country renowned for its wealth and health care infrastructure. Yet the public health infrastructure to support sick bodies during a pandemic did not live up to Switzerland's reputation. Arguably, the Swiss Covid-19 strategy rushed to reopen for economic reasons and additionally failed to harness interrelated community. Geneva-based World Health Organization (WHO) envoy on Covid-19, David Nabarro is quoted as having said that 'Switzerland is among the countries that relaxed too soon and failed to adequately prepare for a second wave', citing a Swiss friend who struggled to get treatment when he fell seriously ill with the virus (Swissinfo 2020).

The vulnerability of health structures to support sick bodies was worsened by the failure to recognize that in this crisis bodies need each other. The lack of emphasis on the social infra-structure that makes interdependence possible was minimized, exposing the neoliberal individualized logic that the body is independent. Covid has laid bare the interdependent nature of human lives and structures. Nabarro's critique of the Swiss response to Covid is instructive in that regard. He noted that –

> The problem lies with inadequate support structures in local communities, which can better give people vital information, monitor the population and act quickly if outbreaks occur. This network should include churches, schools, local com-panies, sports and social groups. (Swissinfo 2020)

The structures mentioned are supported by 'bodies' that are often invisible in the grind of daily life outside a pandemic. These are bodies that are committed to seeking the common good. They often operate in 'dominated occupations where race, class and gender play a significant role in rendering them invisible in normal times' (Clavijo 2020, p. 703). By exposing the vulnerability of those structures, the pandemic has brought to the fore inequalities that are structurally imbedded in the socio-economic infrastructure that supports society. Embodied resistance against such structural injustice entails social justice and promotes the common good. It is an onward, incremental struggle. Ngugi wa Thiong'o captures that aptly 'Our lives (read: bodies) are a battlefield on which is fought a continuous war between the forces that are pledged to confirm our human-ity and those determined to dismantle it' (Ngugi 1987, p. 53).

What justice exists and/or is possible for those exposing themselves to infection to keep us healthy, eating and working – those educating and raising our kids, those keeping our envi-ronment clean and transporting us and our goods? They are usually underpaid, under-appreciated and overworked, and yet they are doing the heavy lifting for the common good dur-ing the pandemic. To what extent am I/are we structurally complicit in the commodification of their vulnerability?

What inspiration will shift our focus from sorely focusing on our personal vulnerability to the vulnerability of others?

A Hermeneutic for Openness

Christianity is fundamentally a religion of the body (see also Chapter 16 by K. Christine Pae). The radical embodiment of God entails divine vulnerability as a condition, resource and language for our own vulnerability. Vulnerability is thus a site for theological, moral and ethical reflection. The search for the protection of one's vulnerable body, family and community opens us up to God and to the structural dimensions of what it means to seek the common good.

The vulnerability of persons and structures calls for a hermeneutic of openness. Through the lens of the Spirit, openness inspires the 'body' of Christ towards a life-giving, truth-telling resistance. It reclaims vulnerability as an asset that profoundly connects our lived realities to God, thereby opening us up to a redemptive encounter. The redemptive encounter is not a triumphal declaration, rather an openness of our human condition to 'both harm and transformation'.

Liberation theologians such as Jon Sobrino and Leonardo Boff have accounted for a vulnerable God through the suffering of Jesus Christ and paralleled Christ's suffering with the suffering of the poor. The insights from liberation theology offer the Christian tradition resources for the critique of structural sin thereby thickening the understanding of human suffering which is mostly understood in terms of divine punitive response to personal sin.

The birth, life, suffering and death on the cross of Jesus Christ who came to save the world is an example of God's vulnerability. Christological debates have wrestled with the question of the seeming helplessness of this redemptive narrative. The Christ who saves the world by his suffering (Matt. 8.17) seems weak and powerless. The suffering power of God's love leaves room for lament – the description of the real situation. This is only one dimension of God's power revealed.

Christianity's claim that God became human, material body is a bold and radical claim. God's self-identification with humans underpins the interconnection of the entire web of life as the 'body of God' (MacFague 1993, p. vii). It links the redemptive love of Christ with the loving creative power of God. Vulnerability and creativity are linked in God. Creator God 'gave birth' (created) all that is and lives in a complex relationship with all of life that is intricately interrelated. The very fact of being related entails vulnerability, which further entails openness to ongoing growth and transformation. The vulnerability of human beings and that of the earth are intricately bound together as the embodiment of God. The dynamic relationship that effects change in the other may be harmful or bad and still be transformative.

The 'bodily' resistance of the earth is evident during the pandemic as reduced human activity has made regeneration of ecosystems possible to some extent. The visibility of service workers who keep the structures that support our lives working have also been highlighted by the pandemic in ways that could portend change towards what the world could become. To that extent vulnerability as a result of Covid-19 puts an accent on ethical and moral obligations we need to take into account in the course of interactions of various vulnerabilities in various aspects of social life. Whether the situation the world is in will definitively lead to heightened compassion and commitment to redress the injustices that these situations have exposed is a complex question. It does not preclude that further vulnerability because the prevalence of shared vulnerability could unleash survival instincts that prey on the vulnerabilities of the weaker.

To be human is to be vulnerable. This basic realization has theological and ethical implications. Being a vulnerable creature entails that a human being needs protection against threats to life. To assume that protection against vulnerability is possible is a legitimate assumption when the possibility exists to surmount it. What power is available to us to protect ourselves and others against vulnerability? What resources, theological or otherwise, do we need to keep moving towards compassion and trust?

Being vulnerable is a condition of human life because human beings will always be vulnerable, not least because vulnerability is inevitable. Yet, vulnerability also makes possible our reflection on, and achievement of humane action. As such, human moral and ethical capacity and accountability arise from vulnerability as the basic human ability to be open to our immediate environment, to other people and other forms of life. With such openness we can reflect on our own pain, recognize the pain of others and (potentially) take responsibility to assuage not only our own pain but also that of other(kind)s. Self-preservation and altruism are hereby brought together as complementary dimensions. This echoes the command to love one's neighbour as oneself (Matt. 22.39) as a two-sided single Christian injunction.

Vulnerability as a Site of Ethical/Moral Reflection

The vulnerability of the poor and powerless in the wake of Covid may intersect with the vulnerability that is part of our human condition, but it is not the full account. The skewed social, political and cultural organization of society means that such vulnerability results from oppression, violence and injustice that are imbedded in societal structures. A world organized around concentration of power and wealth, where domination and violence are the norm, demands agitation for freedom and struggle to change the world order itself.

During Covid, the ensuing deaths have made obvious the more dangerous scourges of discrimination such as racism, labour exploitation and scapegoating, which deny others shared humanity. The environmental challenges which put humanity at risk remind us of the human obsession with our own importance and domination of the 'other'. Examples abound that the people bearing the brunt of the pandemic are the structurally exploited 'others'. As Vinoth Ramachandra reflects in a blog post (2020):

There is no doubt that the global spread of Covid-19 has exposed the lies, hypocrisies and fault-lines that run through many of our societies ... Covid-19 has also exposed how dependent we are on those on the 'underside' of our societies. The people at the frontline of the fight to protect us from the pandemic are the very people whom we routinely ignore, sometimes even revile, and – if the hiTech companies have their way – will soon be replaced by robots: those involved in social care, nurses and hospital orderlies, janitors, sales assistants, garbage collectors, undertakers, mental health workers and migrant labourers on farms and in the food industry.

Covid is a fresh reminder of the universality of human frailty. It has cast an instructive light on vulnerability of the human body and exposed pre-existent social and economic fissures within communities, nations and the globe that are a legacy of falsehoods, duplicities and fault lines that lie between the rich, powerful and privileged few on one hand, and the majority poor and powerless, on the other.

The structural injustices that Covid makes obvious remind us that our ways of inhabiting life need to draw us together rather than tear us apart. The Covid crisis has exposed that vulnerability is as threatening as it is powerful in drawing us together to work for the *common good*.

Vulnerabilities

A theological exploration of vulnerability is understood in terms of mobilizing the 'body' (physical and ecclesial) to move beyond triumphalist piety towards an embodied resistance. The vulnerabilities that Covid-19 has exposed incite/invite us as the vulnerable 'body' of Christ to lament, protest and disrupt as ways of bearing faithful witness to a vulnerable God. The resurrection of Jesus Christ in power gives the body of Christ, the church, new life in the Spirit, here and now. God sends the church into the world in the power of the Spirit. Exercised in vulnerability, that power entails that we are powerless to reclaim lost lives. Nor could we transform every injustice or restore what has been stolen. Faith in a God who shares the suffering of Christ and initiates the power of the Spirit, through his resurrected body is a movement that progressively disrupts, subverts and affirms life, defying the life-denying powers. Resistance against interconnected vulnerabilities is spurred on by the interconnectedness of life itself. The faith that triumphs over vulnerabilities is a profound defiant affirmation that though some die, and though the losses of the past may be unrecoverable, the struggle for life continues. And so, though my body may die the struggle for life does not end. To struggle with hope is to realize that

[though] no mortal lives forever; but life is generational ... living will end, but life passes through; disease will infect, pain and stop the living; but life is not out of touch, and life should not be out of touch ... for all creatures and plants in the sea, under the ground, above the skies and on earth. (See Chapter 12 by Jione Havea)

References

Marilyn McCord Adams, 2006, *Christ and Horrors: The Coherence of Christology*, Cambridge: Cambridge University Press.
Nathalie Clavijo, 2020, 'Reflecting upon Vulnerable and Dependent Bodies during the Covid-19 Crisis', *Feminist Frontiers, Gender Work Organ 2020* 27: 700–704.
Sallie McFague, 1993, *The Body of God: An Ecological Theology*, Augsburg: Fortress.
Toni Morrison, 1987, *Beloved*, New York: Alfred A. Knopf.
Ngugi wa Thiong'o, 1987, *Devil on the Cross*, Portsmouth: Heinemann.
Elizabeth O'Donnell Gandolfo, 2015, *The Power and Vulnerability of Love: A Theological Anthropology*, Minneapolis: Fortress Press.
Vinoth Ramachandra, 2020, 'Who will Learn from Covid-19', 27 April (accessed 8.11.20: https://vinothramachandra.wordpress.com/).
Edward Schillebeeckx, 1980, *Christ: The Experience of Jesus as Lord*, New York: Seabury Press.
Valerie Smith, 2012, *Toni Morrison: Writing the Moral Imagination*, Hoboken: John Wiley & Sons.
Heike Springhart, 2017, 'Exploring Life's Vulnerability: Vulnerability in Vitality' in Heike Springhart & Thomas Günther (eds), *Exploring Vulnerability*, Göttingen: Vandenhoeck & Ruprecht.
Swissinfo, 2020, 'WHO Pandemic Envoy Slams Swiss Covid Response', *Swiss Info*, 21 November (accessed 7.1.21: www.swissinfo.ch/eng/who-pandemic-envoy-slams-swiss-covid-response/46176102).

14

Solidarity Assurance:
Reality, Faith and Action

SITHEMBISO S. ZWANE

Solidarity Assurance is an act of defiance against a pandemic that has caused havoc around the globe. This unprecedented health and economic catastrophe plunged the world into social, economic and political crisis. Developing countries have been hit hard. Churches in developing countries had to learn new ways of being church. Social and ecumenical movements reignited their structures to ensure solidarity with those most affected by Covid. It became urgent to reread the signs of the time and to strengthen solidarity with poor and marginalized communities. Solidarity is critical; it requires progressive dialogue in the community context. In this community context, the working-class poor offer the intellectual capital to engage with the challenges of the pandemic.

First, the reality of Covid exacerbated the problem of an already struggling South African economy that resulted in economic downgrade prior to the lockdown on 26 March 2020. This reality led to the decline in investor confidence and the shedding of mostly non-standard types of work.

Second, the faith dimension suffered as the pandemic divided some church communities between members and the leadership. The poor churches were not spurred in this volatile situation as members and leaders lost income. The ecumenical movement and other churches showed their commitment in prioritizing the poor in an effort to mitigate against the challenges of Covid from a biblical and theological perspective, ensuring solidarity with the poor affected by the situation.

Third, the pragmatic action by the Ujamaa Centre in partnership with the Solidarity Fund movement contributed
immensely to addressing the challenges of food insecurity in the
communities of the poor and the vulnerable in South Africa.
Their vulnerability was not a sign of weakness, but rather an
opportunity to demonstrate the poor's agency against those
responsible for their vulnerability.

Reality: Socio-Economic Impact of Covid on Poor Communities

The African continent has been affected by the Covid pandemic with the majority of people infected and affected
by the virus in South Africa. South Africa has more than
1 million people infected with more than 900,000 recovered
from the pandemic. Unfortunately, more than 33,000 people
succumbed to the pandemic as of 11 January 2021. The good
news is that the recovery rate has increased to more than 90
per cent. The critical challenge facing the continent is the
balancing act between the protection of life and saving the
ailing economy.

The markets in Africa were severely affected by the pandemic. Ozili Petersen reports that the Johannesburg Stock
Exchange (JSE) slumped to 3.7 per cent due to uncertainties
in the markets (Petersen 2020, p. 4). Tourism to South Africa
dropped by 80 per cent following the Covid outbreak and the
situation worsened when the lockdown curfew was imposed
(Petersen, p. 4). Furthermore, according to John Ataguba the
micro-economic costs of the Covid pandemic relates to the
challenges borne by 'individuals, households, firms and other
establishments like schools, hospitals, clinics, health centres,
health facilities, health workers and the government' (Ataguba
2020, pp. 325–6).

The macro-economic impact of Covid was felt with the
retrenchment of workers (Ataguba, p. 327) in different sectors
of the economy. The economic costs of the pandemic have
exacerbated the inequalities between the 'have' and the 'have

nots' in the society. The micro-economic sector has collapsed due to Covid (Ataguba, p. 327).

The Small Micro-Medium Enterprises (SMMEs) was severely compromised because of its informality and fragility. The informal economy characterized by informal traders in the streets was obliterated by the pandemic, especially during the early stages of the South Africa lockdown. These businesses could not survive without income for more than two months. As a result, 20 per cent to 33 per cent of job losses were reported, which is to say that 1 to 1.7 million individuals fell into the poverty trap (Jain *et al.* 2020, p. 2).

Covid will have long term effects on the labour market in South Africa and will further complicate the creation of jobs. The major concern for the majority of poor communities is the availability and accessibility of food. Individuals who lost their work during the lockdown are struggling to provide food for their families while waiting for their payments from the Unemployment Insurance Fund (UIF) and Relief Grants. Reports suggest that 'women have seen a reduction of 49 per cent in active employment over the February-April period in South Africa' and this is 15 per cent greater than for men (Jain *et al.*, p. 8). Women carried the burden of providing for their families during the lockdown when their partners lost their jobs. During this period, different challenges emerged within families.

In South Africa there has been an increase in reported cases of Gender Based Violence (GBV) during the lockdown suggesting that the pressure of lack of income contributed to the physical abuse of women and children by men. Nancy Stiegler and Jean-Pierre Bouchard note two trends: a sharp decline in domestic abuse due to a ban on alcohol, but an increase in domestic violence against women in the third week of the lockdown (Stiegler and Bouchard 2020, p. 2). The increased pressure to provide food for households in the midst of job losses seem to have escalated the problem. Food access for the majority of poor communities was a critical concern in the context of unemployment affecting both women and men.

Joseph Glauber *et al.* argue that 'there will be serious threats to the access of the poor to food as a consequence of lost income from lockdowns and other restrictions' (Glauber *et al.* 2020, p. 1). The restrictions have the propensity to further worsen the already volatile situation contributing to food insecurity and hunger. Glauber *et al.* further argue that 'major exporters and importers of staple food should desist from imposing trade barriers in response to Covid-19 pandemic' (Glauber *et al.*, p. 3). The availability and accessibility of food was important during the lockdown to provide support to the hardest hit poor communities. The challenge of food security severely affected poor communities in rural and urban areas.

In the light of this, the South African government introduced the emergency social insurance or the Relief Social Assistance grants to assist the most vulnerable in the community (Glauber *et al.*, p. 3). However, the social relief grants were not without controversy as the majority of the people who lost their businesses and jobs waited without success for their payouts from their respective employers and the South African Social Service Agency (SASSA). These delays compounded the already fragile economic context of food insecurity in South Africa making households susceptible to hunger.

Glauber *et al.* state that with 'Covid-19 pandemic and its economic fallout spreading in the poorest parts of the world, many more people will become poor and food insecure' (p. 16). In the context of this reality, how did religion or faith mitigate the reality of the Covid-19 pandemic in the South African context?

Faith: Preferential Option for the Poor Communities

Religion or faith in the South African context plays a significant role in the restoration of hope. Furthermore, religion plays an important role in South Africa and beyond especially during the most challenging times (Stuckey 2001, pp. 69–84; West 2008, p. 118). According to John Mbiti, it is 'African religion which gives its followers a sense of security in life' (Mbiti 1975, p. 15). This sense of security serves as a buffer for the poor in times of crisis. Furthermore, Africans believe that misfortune happens with the knowledge or the permission of God (Magesa 1997, p. 41). In what way does God relate to the poor in the time of crisis?

The notion of the 'Preferential Option for the Poor' seeks to 'acknowledge the multifaceted scope of poverty while standing in solidarity with socially insignificant and excluded' (Gutiérrez and Groody 2014, pp. 2–3). Furthermore, this human solidarity is achieved both by engaging in the class struggle as well as by siding with the victims. The Preferential Option for the Poor is not just rhetoric or slogan, but a biblical and theological mandate to liberate the poor from their suffering and to give them abundant life (Ex. 3.7–10; John 10.10, NIV).

It is inconceivable that Christians can be neutral in situations of injustice, because they have a responsibility to side with the poor (Gutiérrez and Muller 2015; Dillon 2013). Our theology teaches us that God through Jesus Christ came to preach good news to the poor (Isa. 61.1–2; Luke 4.18–21). According to liberation theology, the poor receive special respect because they are 'especially situated to hear the Word of God' (Tambasco 1986, pp. 38–40).

The 'preferential' option in the liberation theology narrative is concerned about the insignificant in the community. The good news of liberty from captivity and release from indebtedness is primarily about the option for the poor. According to Kirylo James, poverty is comprehensive and reflect individuals who lack the economic means to sustain themselves in a healthy and productive existence (Kirylo 2006, p. 267).

The most contested term in the title is 'preferential', which connotes some kind of 'preference' or a 'choice to be in solidarity' with those affected by poverty. The critique of the term includes that 'preference' violates agape, particularly the love of enemies when it depicts the 'nonpoor' as 'class enemies' (Pope 1993, p. 243) who must be defeated rather than loved. The Preferential Option for the Poor relates to issues of proper or improper partiality. But it gives the poor power to speak prophetically and theologically against their oppressors. It provides the poor with a platform to mobilize against the oppressors who perceive them as vulnerable.

The God of the Bible is against suffering (Ex. 3.7–10), hence the importance of liberation theology as praxis for the poor and the oppressed. Liberation theology emerges from an act of compassion by God for those who suffer in the midst of crisis. Segundo asserts: 'I do not believe that there is any other way of expressing the option for the poor concretely than to say it is God's compassion for the most afflicted' (Segundo 1993, p. 125). The grasp of the concept of compassion is critical in the context of suffering and marginalization. In this context, liberation theology requires prophetic faith and action to foster the progress of the oppressed or marginalized in the community (Segundo, p. 125).

The understanding and practice of compassion is the prerequisite to the notion of 'preferential option for the poor' (Kirylo 2006, p. 268). 'Compassion' comes from the Latin word *compati* which refers to being conscious and aware (of your neighbours' challenges and difficulties) (Kirylo, p. 268). The church has the biblical responsibility to show compassion and be merciful to the poor and the vulnerable during and after the Covid pandemic. This is the Preferential Option for the Poor premised on the good news for the poor and the afflicted.

The South African Council of Churches (SACC) acknowledges in a pastoral letter the dire situation that the poor communities face because of Covid. The Council takes the prophetic stance to side with poor communities. The SACC takes the Preferential Option for the Poor who are hungry, desperate and frustrated by unprecedented disruption to their lives. Furthermore, the SACC's pastoral letter underscores that Covid has laid bare the effects of poverty and inequality in South Africa (SACC 2020, p. 2). The SACC reaffirms that the church is the vehicle to articulate a theology of liberation to bring hope to the hopeless and liberty to the oppressed and suffering.

The SACC states that 'to be the light of the world requires a living consciousness of developments and going on in the world and society, in which we discern the "signs of the times" so that we can be a viable and helpful "light of the world"' (SACC, p. 2). The fundamental message is that the church is home to those who are struggling and suffering. It is also strategically placed to engage in praxis, to read 'the signs of the times' by analysing the social, economic, religious and political 'signs' in order to develop an alternative redemptive theological narrative that identifies with the poor communities affected by Covid in South Africa and beyond.

The SACC states that 'God wants us to proclaim a message of faith' (SACC, p. 2), a message of abundant life (John 10.10) in the midst of suffering and death. The SACC is unequivocally clear about their role in addressing the impact of Covid: 'we must adopt a public ministry that best serves and protects the weakest among us; in this case the most at risk from the Coronavirus, for God's sake' (SACC, p. 2). The scriptures remind us of the importance of being in solidarity with those in need of help and support, in the times of need (Matt. 25.40). The SACC further states that Jesus Christ calls on the followers to read 'the signs of the times' (Matt. 16.3). The 'signs' of our time are appropriated through the biblical and theological 'voice' that engages directly with scriptures (and the church) as a site of struggle in search of an alternative theological narrative that deconstructs the structures and systems of oppression (Zwane 2020b, 9). This is a public prophetic church conscious of the socio-economic ills in the community.

Public prophetic theology is premised on the idea of 'reading the signs' (Vellem 2010) of our time as stated by the SACC and speaking theologically and prophetically to crisis situation. It is time to 'read the signs' of our contemporary time once again in the midst of Covid, HIV and AIDS, Gender based Violence, unemployment, poverty, inequality and land access. The SACC in its general pastoral letter from the leaders of the member churches stipulated: 'We address ourselves to the anxieties of our people, the hunger, the desperation, the frustrations; and the depression that is setting in' (SACC, pp. 1–2).

The SACC in this statement postulates that a public prophetic church should be concerned when the poor suffer. The ecumenical church in the form of Faith Based Organizations (FBOs) and social movements collaborated with Non-Governmental Organizations (NGOs) to support vulnerable poor communities and churches in KwaZulu-Natal. The ecumenical movements and other churches demonstrated commitment in prioritizing the poor in an effort to mitigate against the challenges of Covid The pertinent question is: What does it mean to do theology in the time of Covid? Jesus articulates what constitutes solidarity theology, stating that what you do to the least of these you have done to me (Matt. 25.34–46). The Covid pandemic somewhat limited the church's ability to do solidarity theology as articulated in Matt. 25.34–46. Jesus in this text is apprehensive about caring for the most vulnerable in society during the time of crisis. The unprecedented physical distancing affected the church's routine ministry of physical presence. However, work continued, and the ecumenical church showed solidarity by doing what Jesus commanded the church to do.

The Ujamaa Centre for Biblical and Theological Community Development and Research at the University of KwaZulu-Natal organized a series of community-based training on Covid with the affected poor communities, the church leaders from mainstream (historical), Charismatic and African Independent Churches (AICs) as part of its commitment to a redemptive theological narrative. Ujamaa had three objectives: First, the fundamental objective was to assure the affected poor communities and churches of solidarity in the time of crisis. Second, to determine how these poor communities and churches were affected by the pandemic. Third, to provide much needed food security as part of the pragmatic action as a response to the pandemic.

Action: The Ujamaa Centre's Response to the Affected Poor Communities

The Ujamaa Centre's methodology of See–Judge–Act (West and Zwane 2020, p. 7) laid the foundation for the discussion on the impact of Covid on the lives of poor communities in the rural areas of KwaZulu-Natal. The Ujamaa Centre's work on Contextual Bible Study (CBS) (West 2016; West and Zengele 2006; West and Zondi-Mabizela 2004; West and Zwane 2013;

West and Ujamaa Centre 2014) is part of the process of reading the 'signs of times'.

The CBS follows a logical sequence of liberation praxis, which is *action* and *reflection*. This is a dialogical process of reflection. In the context of Covid, the Lord's Prayer (in Matthew) was used to engage in discussion with the church leaders affected by the pandemic in the area of Pietermaritzburg, KwaZulu-Natal, South Africa. The CBS on Matthew 6.9–13 (NIV) focused on food security and debts during covid time.

The questions were (see West and Ujamaa Centre, 2014 and 2019):

1 What is the text about?
2 According to Matthew's version of the prayer, where will God's kingdom come?
3 How do verses 11 and 12 describe the kingdom of God?
4 Why is Jesus so concerned about 'bread for today' (verse 11)?
5 Why is Jesus so concerned about 'debt' (verse 12)?
6 What do you plan to do in your church as result of this CBS?

The CBS line of questioning allows the rereading of the biblical narrative to focus on the literary 'detail' of the text. The first question – What is the text about? – is a community consciousness question (West and Zwane, 2020) that is a prerequisite for literary textual analysis.

In the literary textual analysis of the Lord's Prayer, the groups focused on verses 11 and 12 interrogating the challenges of food security and debt affecting the communities during the lockdown as a result of Covid. The participants shared their frustrations over losing their jobs resulting in food insecurity among households, and many have got into debt trying to mitigate against hunger and poverty.

The church leaders said, 'We are suffering, we have no food and we are in debt, how do we minister to people when we are broken?' The leadership lamented their loss of income and access to food, resulting in debt. The members of the church lamented, 'we gave to the church when we had income, but the church is not able to help us during this crisis'. These two contrasting statements indicates polarization within the church as a result of Covid. The importance of solidarity within the church is therefore critical.

The pragmatic question was, what would be the appropriate action to take after the CBS? It was evident that this was a time of hardship and crisis. It was a time of darkness, and the SACC pastoral letter argued for the church to be the 'light of the world' (Matt. 5.14 NIV; SACC, pp. 5–6) and to shine upon those affected in the time of crisis and despair. The Covid pandemic was a time of crisis that required a prophetic public church to 'read the signs of the time' and to be the light of the world in the midst of darkness and death.

The SACC postulated revisiting public prophetic theology that engages contemporary socio-economic and political challenges affecting the most susceptible members of society. The participants in the Ujamaa workshops lamented their inability to mitigate against the challenges presented by the pandemic because of their lack of financial resources. Most of the participants were women between the ages of 25 and 50 working as informal traders selling food on the streets and taxi rank. They explained that, unlike HIV, the Covid pandemic deprived them of their dignity when they lost their jobs as informal traders (*See*). This put a strain on their family life and daily routine because they could not go to work under lockdown, children could not go to school and families could not support each other during the pandemic because of the curfews imposed by the National State of Disaster Act.

The second phase focused on faith (*Judge*) with mostly church leaders between the ages of 30 and 65, whose response was in general positive. The analysis of contested 'spaces' of power (West and Zwane 2020, pp. 17–18) created a platform for further engagement with the scriptures. These spaces reflect the contested terrain that is religion, hence the need to interrogate dominant and oppressive religious 'spaces' (West and Zwane 2020, pp. 4–8) that undermine the contribution of the poor communities. The key question about the impact of

Covid on their lives as church leaders led to some interesting responses: How did the pandemic affect you personally? One responded: 'No salary, great fear – no visitors/or ability to visit others – not sure what to do without members of the congregation, not working in the church, just sitting at home watching the news all day long.'

It is evident from this response that many of the church leaders were severely affected financially, socially and spiritually by the pandemic. They lost income as well as social interaction with their congregants especially in the beginning of the lockdown as people were trying to innovate about methods of communication using online or virtual platforms. Due to lack of income, most church leaders got into debt because they could no longer rely on their members who were also affected by the pandemic. The difference between HIV and Covid experiences is that with HIV the financial and social aspects of life were not severely affected the way that Covid has done (see also Chapter 9 by Beverley Haddad). During early HIV and AIDS infections, life continued as 'normal' on the business side, and it also did not affect social relations.

The third phase of the Ujamaa work focused on the pragmatic response (*Act*) with the actual distribution of food parcels to all the participants that were affected by Covid and their communities. The distribution of food parcels was not an event but an ongoing process of solidarity with the poor communities affected by the pandemic under lockdown. The Ujamaa theory of change maintains that the dominant religious narratives that undermine prophetic theological paradigm cannot be left alone (West 2016, pp. 136–43) while development takes place. Development without the participation of communities is counter-productive (Zwane 2020a, pp. 212–33). Covid changed the 'normal' in our daily routine and accelerated the relevance of the fourth Industrial Revolution (IR) in the twentieth century.

Conclusion

The unprecedented Covid pandemic raised critical questions about the abilities of nations and churches to handle crisis. The priorities are critical in responding to the crisis in terms of who receive care and support. The South African government demonstrated its commitment to caring and supporting the most vulnerable in the community with the provision of social relief funds for the poorest members of the society.

The Ujamaa Centre is committed to ensuring that poor communities have access to justice. Covid demonstrated the social and economic inequalities that exist between communities. It is unequivocally evident that the pandemic does not discriminate with respect to infection, but it does discriminate on the basis of class. The socio-economic impact of the pandemic was mostly felt among the working-class poor communities. The Covid pandemic revitalized the notion of the 'preferential option for the poor'. The church has been reminded of the importance of solidarity with the poor and the afflicted in the time of crisis.

The response of the Ujamaa Centre is reflected in the three sections of this chapter, drawn from the See–Judge–Act of liberation theology. First – *See* – this chapter underscores that the new 'normal' of Covid caused heartache for many people around the globe. The South African government is apprehensive that Covid has exacerbated the problem of an already struggling economy that resulted in the downgrade prior to the lockdown on 26 March 2020. The sharp decline in the markets meant that more people lost their jobs and are part of the growing list of the unemployed. Covid exposed the realities of economic inequalities in society.

Second – *Judge* – is the importance of faith and the notion of the preferential option for the poor in caring and supporting poor communities in our context. The poor churches were not spurred in this volatile situation especially the working class. The ecumenical movements and other churches showed commitment in collaborating with the poor in an effort to mitigate the challenges of Covid.

Third – *Act* – the pragmatic action by the Ujamaa Centre in partnership with the Solidarity Fund movement contributed immensely in addressing the challenges of food insecurity in the poor communities in South Africa. The Contextual Bible Studies (CBS) with a range of stakeholders in the community resuscitated hope among the leaders in the community. The solidarity with the poor and the marginalized was reactivated as a result of the Covid pandemic.

References

John E. Ataguba, 2020, 'Covid-19 Pandemic, a War to be Won: Understanding its economic implications for Africa', *Springer nature Switzerland* 325–8 (doi:10.1007/s40258-020-00580-x).

Eilish Dillon, 2013, 'What role do faith-based values play in the development process and in wider social and economic change in developing countries?' Dublin: Kimmage Development Studies Centre, All Hallows College.

Joseph Glauber, David Laborde, Will Martin and Rob Vos, 2020, 'Covid-19: Trade Restrictions are worst possible response to safeguard food security', IFPRI Blog (accessed 20.3.21: www.ifpri.org/blog/covid-19-trade-restrictions-are-worst-possible-response-safeguard-food-security).

Gustavo Gutiérrez and Daniel G. Groody, 2014, 'Introduction' in Daniel G. Groody and Gustavo Gutiérrez (eds), *The Preferential Option for the Poor beyond Theology*, pp. 1–8, Notre Dame: University of Notre Dame University Press.

——— and Gerhard Ludwig Muller, 2015, *On the Side of the Poor: The Theology of Liberation*, Maryknoll: Orbis.

Ronak Jain, Joshua Buddlender, Rocco Zizzamia and Ihsaan Bassier, 2020, 'The labor market and poverty impacts of Covid-19 in South Africa', Cape Town: *SALDRU* working paper No. 264.

James D. Kirylo, 2006, 'Preferential Option for the Poor, Making a Pedagogical Choice', *Childhood Education* 82.5: 266–70 (doi:10.1080/00094056.2006.10522839).

Laurenti Magesa, 1997, *African Religion: The Moral Traditions of Abundant Life*, Maryknoll, New York: Orbis Books.

John S. Mbiti, 1975, *Introduction to African Religion*, 2nd edition, London: Heinemann Educational Publishers.

K. Ozil Petersen, 2020, 'Covid-19 in Africa: Socio-Economic impact, Policy Response and Opportunities', *MPRA* paper no 99617 (accessed 20.3.21: https://mpra.ub.uni-muenchen.de/99617).

Stephen J. Pope, 1993, 'Proper and Improper Partiality and the Preferential Option for the Poor', *Theological Studies* 53, Boston: Boston College.

John Luis Segundo, 1993, *Signs of the Times*, Maryknoll: Orbis.

South African Council of Churches (SACC), 2020, 'A General Pastoral Letter from the leaders of SACC Member Churches in Time of Covid', Johannesburg.

Nancy Stiegler and Jean Pierre Bouchard, 2020, 'South Africa: Challenges and Successes of the Covid-19 Lockdown', *PubMed.gov* (doi:10.1016/j.amp.2020.05.006).

Jon Stuckey, 2001, 'Blessed Assurance: The role of religion and spirituality in Alzheimer's disease caregiving and other significant life events' (doi:10.1016/s0890-4065/(00)00017-7).

Anthony J. Tambasco, 1986, 'Option for the Poor', in *The Deeper Meaning of Economic Life*, R. Bruce Douglas (ed), pp. 37–55, Washington DC: Georgetown University Press.

Ujamaa CBS Manual Part 2, 2019, 'Matthew 6: 9–13: Land and Food Security', 87–8 (www.ujamaa.ukzn.ac.za).

Vuyani S. Vellem, 2010, 'Prophetic Theology in Black Theology, with special reference to the *Kairos document*', *HTS Teologiese Studies / Theological Studies* 66.1: 6 pages (doi:10.4102/hts.v66i1.800).

Gerald O. West, 2008, 'The ANC's deployment of religion in nation-building: from Thabo Mbeki, to "The RDP of the Soul", to Jacob Zuma, in Thabo Mbeki's Bible: The role of religion in the South African public realm after liberation.' *Bulletin for Old Testament Studies in Africa*.

———, 2016, 'Recovering the biblical Story of Tamar: Training for Transformation; doing Development', in *For Better for Worse*, Robert Oden (ed.), Swedish Mission Council, pp. 136–42.

——— and Sithembiso S. Zwane, 2020, 'Re-reading 1 Kings 21:1–16 between Community Based Activism and University-Based Pedagogy', *Journal for Interdisciplinary Biblical Studies Activism in the Biblical Studies Classroom: Global Perspectives*.

——— and the Ujamaa Centre for Biblical and Theological Community Development and Research, 2014, *Doing Contextual Bible Study: A Resource Manual*, Pietermaritzburg: The Ujamaa Centre for Biblical and Theological Community Development and Research.

——— and the Ujamaa Centre for Biblical and Theological Community Development and Research, 2019, *Doing Contextual Bible Study: A Resource Manual*, Pietermaritzburg: The Ujamaa Centre for Biblical and Theological Community Development and Research.

——— and Bongi Zengele, 2006, 'The Medicine of God's Word: What people living with HIV and AIDS want (and get) from the Bible.' *Journal of Theology for Southern Africa* 125: 51–63.

―――― and Phumzile Zondi-Mabizela, 2004, 'The Bible story that became a Campaign: the Tamar Campaign in South Africa (and beyond)', *Ministerial Formation* 103: 4–12.

―――― and Sithembiso S. Zwane, 2013, 'Why are you sitting there? Reading Matthew 20: 1–16 in the context of casual workers in Pietermaritzburg, South Africa', in *Matthew: Texts@Contexts*, edited by Nicole Wilkinson Duran and James Grimshaw, pp. 175–88, Minneapolis: Fortress Press.

Sithembiso S. Zwane, 2020a, 'Invited, Invigorated and Invented spaces: Re-appropriating development approach to Community Participation', in *Faith, Class, and Labor*, edited by Jin Choi and Joerg Rieger, pp. 212–33, New York, Wipf and Stock Publishers.

――――, 2020b, 'Transition, Reflection, Re-thinking and Re-imagining: The relevance of Black Liberation Theology (BLT) post 1994', *HTS Teologiese Studies/Theological Studies* 76.3 (doi:10.4102/hts. v76i3.6078).

15

out of hand

Curves rise as the novel coronavirus mutates
from 19 to 20 and to more years to come
over land and sea, bodies are infected
by contact with loved ones and strangers
with caregivers and careless helpers, many die
in public places, without their normal companions
many more black, brown and minoritized bodies get sick
and die more so than white privileged bodies even though
color and race make no difference at the graveyard
gender, wealth and sexuality make no burning difference in
  the furnace

Parents and grandparents pass away
sons and daughters, grandsons and granddaughters
give their last rites to health workers
and are mourned by grave diggers

The first wave mangled like a rip and pushed like a tsunami
that will not settle back before the second wave breaks
ripping and rippling, with the pangs of labour but
instead of newborns, victims fall
victims who have seen light
experienced injustice and all that
shit of civil societies

Empires on and off shores
prey for daily bread with folded arms
deposit the poor in their pocketbooks
and reopen the old normal
to reboot their estates and balance sheets

Empires roll back environmental protection
invest with funds not yet stolen, look the other way
expecting the infected and dead to be Samaritans
migrants, refugees and sex workers whom civil societies
require and condemn at the same time because
they are little less than toilet paper
essential, but disposable

Has the world gone mad and out of hand?
has the world gone mad in our hands?

Jione Havea
Wurundjeri land & waters
Kulin nations
1 November 2020

PART 3

In Decencies

16

Indecent Resurgence: God's Solidarity against the Gendered War on Covid

KEUN-JOO CHRISTINE PAE

In response to the Covid-19 outbreak, many world leaders pro-voked war languages as if the entire world fought a common war against the virus. The media evoked the images of an apocalyptic war by digitizing dead bodies and spectacularly displaying corpses in plastic bags, medical workers' faces covered with tears, and the empty eyes of people who had lost family members to the virus. These images correspond to Achille Mbembe's concept of 'necropolitics' or sovereignty's right to kill: sovereignty manifests its power by killing those who are condemned to death for the sake of fostering other people's lives (Mbembe 2019). Although many countries' policies to stop the spread of Covid-19 appear to be an exer-cise of 'biopolitics' or allowing people to live, it may not be an exaggeration to say that the neoliberal capitalist world has led Covid-19 to infect and kill vulnerable, invisible, or 'surplus' populations selectively. The war against Covid-19 is, in fact, the war against the global poor, and this war is profoundly gendered and sexualized.

If Covid-19 is considered a form of war, whose bodies are subscribed to death? How do they resist death? What kind of theological praxis would we need to resist the power of death, spread by the covid war rhetoric? I delineate these questions, based on Marcella Althaus-Reid's insightful idea that 'all political theories are sexual theories with theological frames of support'

(Althaus-Reid 2000, p. 176). More specifically, through the lens of gender and sexuality, I analyse the covid time with attention to militarized and masculinized metaphors and practices by political powers. The critical analysis leads to the radically reimagined possibility of transnational solidarity for a new just world order in the post-Covid era. I propose this possibility as transnational feminist praxis for global justice. Since Covid-19 is a global pandemic, we need a transnational approach.

In terms of transnational feminism, this chapter employs M. Jacqui Alexander and Chandra Mohanty's definition of the transnational as:

> (1) a way of thinking about women in similar contexts across the world; (2) an understanding of a set of unequal relationships among peoples; and (3) a consideration of the term 'international' in relation to an analysis of economic, political, and ideological processes that would therefore require taking critical antiracist, anticapitalist positions that would make feminist solidarity work possible. (Alexander and Mohanty 2010, p. 24)

In the covid time, transnational feminist solidarity does not mean to fight against the virus per se but to dismantle oppressive social structures constructed upon white heteropatriarchal capitalist desires that enlarge unequal relationships among peoples. Race, gender, class, sexuality and jurisprudence signify power relations and systematize sovereignty's right to kill or necropower. Hence, the covid time requires us, who are deeply concerned about global peace and justice, to critically interrogate the relations of ruling and the spatiality of power, because one's social identity and geographic location often become determining factors of whether they have life or death. As politicians provoke war languages, we need a transnational feminist materialist analysis of who becomes neglected, intentionally ignored, and forced to care for others. Subsequently, this analysis should unpack how the death and debilitation of the poor during the pandemic are profoundly gendered and sexualized. For transnational feminist solidarity to rebuild a new world order, I borrow wisdom from Marcella Althaus-Reid's *Indecent Theology*. This chapter will propose 'indecent resurgence' as a counter-narrative to the war on Covid-19 and as a path for transnational solidarity for justice.

War on Covid-19: War Against the Poor

War metaphors for the pandemic are compelling enough to get people's attention to the danger quickly. On 5 April 2020, Queen Elizabeth ended her unusual speech with the phrase 'we will meet again'. Her speech evoked the famous 1939 Second World War ballad, 'We'll Meet Again' by British singer Vera Lynn, as if the UK had entered another world war. Donald Trump called himself the war president. War metaphors were accompanied by urgent sovereign decisions to mobilize nurses and doctors as first responders on the frontline, shut down schools, ban travels, lockdown cities, and impose various social and health regulations on its people.

When the whole world fights for the same war, its interest goes to who wins the war and how they win. The death toll measures a country's success in winning the war. However, war metaphors for Covid-19 are misleading and even dangerous. The metaphors encourage people to look for embodied enemies, erase victims' real faces by digitizing their deaths, reinforce patriarchal gender ideology that glorifies an ethic of care, and concentrate medical resources on treating the virus. Poor people's lives become more precarious during the pandemic just as any war destroys the poor's lives first.

On a surface level, the war on Covid-19 portrays the virus as an enemy. However, as right-wing politicians and media worldwide occasionally call it the China or Wuhan virus, the enemy begins to have an East Asian face. In the United States, anti-Asian attacks were tripled at the beginning of the nationwide lockdown, compared to pre-Covid time (Rogin and Nawaz 2020). Asian women are more vulnerable to anti-Asian violence, just as the war on terrorism aggravated anti-Muslim violence, especially targeting Muslim women. Asians or anyone who is perceived as a virus carrier or free-rider on the health-care system can be enemies to society. The war metaphor frequently divides society into 'us (read as innocent victims to the virus or fighting soldiers)' and 'them (read as a threat to public health and social stability)'. In this war on Covid-19, people become either soldiers fighting for public safety or enemies to the public – no one can be a bystander or a civilian.

In the meantime, as South Korean researchers argue, the war metaphor justifies the exploitation of civil servants and first respondents (Kim *et al.* 2020). As a result, moral injury and post-traumatic stress disorder of these healthcare workers will become inevitable. Both moral injury and PTSD are found among combat veterans, while war fatigue or numbness to war is common among civilians. Moral injury is the concept, first developed by psychiatrist Jonathan Shay, who has treated Vietnam War veterans. According to Shay (2014), moral injury occurs when there has been a betrayal of what is right either by a person in legitimate authority or by oneself often in a high stakes situation such as combat. Moral injury results in deep guilt, shame, anger, or distrust in one's moral capacity. As morally injured people cannot trust their ability to make morally sound decisions, their sense of humanity dissipates. Addiction to alcohol and drugs becomes a coping mechanism for them. They may express uncontrollable anger and use vio-lence against other people, including their families. In the worst case, they may kill themselves and others. Although the studies of moral injury concentrate on combat veterans, healthcare workers, who are now considered fighters on the frontline, work in morally injurious situations where they have to make 'difficult decisions related to life and death triage or resource allocation', or believe that 'a patient's life might have been saved under different circumstances' (Watson *et al.* 2020). The workers may also feel betrayed when they witness 'what they perceive to be unjustifiable or unfair acts of policy' and guilty about surviving when others are dying or for infecting people who are close to them (Norman and Maguen 2020).

Furthermore, the dichotomous rhetoric of 'us' and 'them' causes the surge of nationalism across the globe. Although the impact of Covid-19 is transnational, sovereign nations rely on patriotic nationalism to combat the virus. Many countries have been busy securing their borders from outsiders. They chose isolationism over collaboration. The best example of collectivized self-interested nationalism is the competition over Covid vaccines, life-saving ventilators, and other essential medical supplies that treat those with severe symptoms. The Covid war victimizes third-world countries that cannot participate in this competition in the neoliberal global medical supplies market. The global poor, whom neoliberal capitalism has considered 'surplus' and thus disposable, suffer the most in this war.

Finally, the war rhetoric 'we are in this together' erases the real faces of the dying. The global poor who are physically and metaphorically dying from the virus are predominantly people of colour, Indigenous people, prostitutes, migrant labourers, domestic workers, homeless people, undocumented immigrants, refugees, people with disabilities, and poor older people.[1] Poor children and youth are faceless and voiceless victims, too. The virus is not the only cause of Covid death and the debilitation of the global poor; the neoliberal global market economy constructed upon white heteronormative patriarchal family ideologies is also causing death.

Women, Gender, Sexuality and Covid-19

Who are the global poor exponentially affected by Covid-19 due to inequality? If analysing what happens to the poor is the essential step for antiwar, anticapitalist and antiracist theological discourse, seriously considering how to approach the poor's experiences without idealizing them should precede this analysis. This caution is precisely concerned about the ethics of knowledge production and critical awareness of the politics of spatiality, as defined by Alexander and Mohanty (2010). Here, Marcella Althaus-Reid's method for indecent theology is a useful tool.

According to Althaus-Reid (2000), Latin American liberation theology began as an indecent theology because its praxis attempted to dismantle colonialist theology. However, it turned decent as liberation theologians compromised their methods and God-talk to heteropatriarchal systematic theology (gender and sexuality-neutral discourse) to attract larger audiences in the neoliberal market. Hence, it is crucial for liberation theologians to see the gendered and sexualized realities of the poor and to reflect theologically on the sexual stories of the poor exploited by the neoliberal capitalist. The sexual stories from the poor inevitably generate what Althaus-Reid calls Indecent Theology, the sexual understandings of God, Jesus and Mary, the mother of Jesus. Indecent theologians are called to be 'sexual performers of a committed praxis of social justice and transformation of the structures of economic and sexual oppression in their societies' (2000, p. i). Althaus-Reid challenges theologians to analyse heteropatriarchal family values which are the root of the 'decent' political-economic structure of globalization. At times of crises like war and Covid, heteropatriarchal family values are reinforced while political authorities convince the public to find solutions to crises within the militarized capitalist system. For this chapter, I examine an ethic of care's violence towards women and gender-based violence, the two co-constitutive violences that are quickly escalated by militarism.

Ethic of Care

In the so-called time of peace, an ethic of care is considered decent. Compassionate care which arises in the understanding of a self in relation with others is a noble value which Christian ethics justifies with languages of love, justice and accountability. However, when crises reinforce heteropatriarchal family values, an ethic of care may become violent. During the covid time, women have experienced (structural) violence of care at various levels at workplaces and homes. Women predominate in workforces at caring facilities, including hospitals and nursing homes, where group infections often occur. Hospitality industries such as hotels, restaurants, domestic labour and childcare services, severely hit by Covid-19, also employ predominantly female workers. At home, the burden of caring falls mostly on women's shoulders. Home-schooling, preparing meals for families, who now spend most time at home, and caring for the sick and the elderly add to many women's usual household labour.

Women's care work has been undervalued because it is not considered a real economy. Now in the neoliberal capitalist world, many women work both inside and outside the home to make ends meet. Women who work outside the home, for example academics, are exhausted and frustrated as their research and academic productivity are halted while their domestic responsibilities increase. Almost seven months after the Covid-19 outbreak, the United States overall has regained nearly half of the lost jobs. However, mothers of school-age children, black men, black women, Hispanic men, Asian Americans, younger Americans (ages 25 to 34), and people without college degrees recover their jobs at a slower rate. Significantly black women recovered only 34 per cent. Mothers of children aged 6 to 12 have recovered less than 45 per cent of jobs lost, while the employment of fathers of children the same age is 70 per cent back (Long 2020). Working at home and attending to their families' needs, many mothers juggle with multiple responsibilities.

Violence

Domestic and family violence, gender violence and femicide have increased across the globe since the outbreak of Covid-19. Many feminist scholars and activists have called gender-based violence a 'pandemic' (the oldest pandemic). According to one piece of research, in Argentina, where the rate of femicide is extraordinarily high during regular times, 86 femicides have already been perpetrated since the beginning of 2020, of which 24 occurred during the Covid plague. In Turkey, 18 women have been killed since the lockdown was instituted, most in their homes (Weil 2020).

However, the masculinized war metaphor for Covid prioritizes eliminating the virus and minimizing the spread of infection. It ignores all other societal problems, including domestic and gender violence. Women's reproductive rights and sexual health, including safe abortion and access to contraceptives, pregnancy and child delivery, prenatal and post-partum care and routine check-ups to prevent breast cancer and cervical cancer are all limited by Covid policies.

Sex workers, either prostitutes or those who work in diverse sex industries, are victimized in Covid time. With Covid, precarity and invisibility of their life have escalated at least in three major areas: (1) many countries where sex work is considered illegitimate, sex workers are excluded from their governments' Covid relief funds and struggling with economic hardship; (2) social stigma and surveillance on sex workers increase as if they were responsible for spreading Covid-19; and (3) lack of access to personal protection equipment (PPE), healthcare for sexually transmitted diseases, and a social safety net (e.g., house and food security and legal status especially for migrant workers) has become exacerbated (Singer *et al.* 2020). Since studies of sex work should be contextualized, sex workers may experience these challenges differently across the globe. However, almost everywhere Covid-19 threatens the lives of sex workers more than before, whether they work or not.

Necropolitical Labour

Feminist scholar Jin-kyung Lee calls sex work 'necropolitical labour' as the most disposable form of labour. The commodified body and acts of a prostitute that simultaneously erase the prostitute's subjectivity is 'one of the most disposable [labour] commodities' (Lee 2010, p. 7). Necropolitical labour constantly exposes labourers to verbal, psychological and physical violence, injuries, death and trauma. At the same time, their life is dependent on this deadly labour for wages necessary for food and shelter. These labourers experience slow dying or living death while their labour is extracted from their bodies for the fostering of others' lives. Prostitution in globalized contexts (i.e., military prostitution, sex tourism and pornography industries) shows unequally constructed interracial, interclass, and intergender relations of ruling. Poor women's sexualized labour fosters not only the lives (or pleasure) of their clients but also those of their dependants and on a national and international level, the economic development of their countries. For instance, sex tourism and pornography industries in Southeast Asia display a one-sided symbiotic relationship between poor women and girls' sexualized labour and national economic development (for the case of South Korea in the 1960s and 1970s, see Pae 2020).

Lee argues that the notion of necropolitical labour demonstrates how the economy of biopower 'operates by destroying and harming the lives of others' (Lee, p. 8). In other words, biopower manipulates and manages death in the form of necropolitical labour, which is 'always already a constitutive dimension of biopower' (Lee, p. 8).

Just as Althaus-Reid accentuates, Lee's conceptualization of necropolitical labour highlights the necessity to analyse proletarian sexualities to unpack the realities of the poor in the neoliberal market economy. Necropolitical labour unapologetically shows raw violence of 'fostering the life of others' which can manipulate an ethic of care, particularly against poor women, because they have historically been forced to respond to others' sexual needs. It should be noted that poor women's sexualized realities seen in prostitution industries are not the results of their personal choices or sins but have been systematized for an indefinite time. As an analytical tool, the notion of necropolitical labour critically interrogates the proletarianization of sexuality that goes with the art of condemning the poor's sexual moralities. What is perceived as immoral or dirty is ready for sexual and material exploitation by those who hold heteropatriarchal power in the neoliberal market. Moreover, if necropolitical labour is constitutive with biopower, political authorities' policies to stop the spread Covid-19 are not biopolitical but always assume some people's necropolitical labour.

While writing this chapter, I listened to Vera Lynn's 'We'l
Meet Again' a few times. The lyrics of the song and Lynn's
husky voice are the gendered nostalgia for the Second World
War that still lives in the collective psyche of people: while
manly men fight the war, good women (mothers, wives, girl-
friends, daughters) wait for them, responsibly taking care of all
domestic matters. We hardly see these women, while all other
women and non-manly men (bad women, sexual minorities,
people with disabilities) are invisible.

Indecent Resurgence for Post-Covid Era

Covid-19 is violent not only because the virus is deadly but also
because political responses to it strengthen heteronormative
nationalist space for racialized, gendered and sexualized violence that kills and debilitates the marginalized at a fast pace.
How can people of faith radically reimagine a God-talk that
has liberative power from death, debilitation and fear? What
we need is not a surge of militant nationalism but a 'radical
resurgence' of Jesus' solidarity movement with sensitivity to
gender and sexuality, intersected with race and class. A radical
resurgence of the Jesus movement is my theologically indecent
proposal for transnational feminist solidarity, overcoming
Covid-19's death-bound social practices and metaphors.

I borrow the term 'radical resurgence' from Nishnaabeg poet,
scholar and activist Leanne Betasamosake Simpson from First
Nation, Canada. Radical resurgence requires 'a deeply critical reading of settler colonialism and Indigenous response to
the current relationship between Indigenous peoples and the
state' (Simpson 2017, p. 48). Adding 'radical' to a resurgence
movement, Simpson emphasizes the importance of taking
back resurgence from neoliberalism and at the same time,
confirming body sovereignty of Indigenous people which the
white settler colonial government of Canada has severely
destroyed, raped, exploited and killed. Resurgence is not about
celebrating cultures dismantled by settler colonialism or merely
retrieving what has been lost in history. Instead, radical resur-

gence engages 'visioning, thinking, acting, and mobilizing' around Indigenous systemic alternatives that respect ancestors, Two-spirit people, non-binary gender hierarchy, nature, and non-human nations' (Simpson, p. 49). Gender must be centred in resurgence because gender violence has been the deadly tool of white settler colonialism.

Indigenous knowledge for a radical resurgence movement teaches interrelationality or interconnectedness among all beings – ancestors, sovereign nations of the natural world (e.g., trees, animals, air, water and earth), tribal nations, people in revolutionary movements against neo/colonialism, heteronormative patriarchy and neoliberal capitalism. In this understanding of interconnectedness, Indigenous activists in radical resurgence pursue solidarity with those who are in other resurgence movements such as the Black Lives Matter movement, black feminism, and other decolonizing movements across the globe. Simpson's proposal for a radical resurgence resonates with transnational feminism. Transnational feminists respect local knowledge while diverse forms of localized knowledge synergize one another and radically reformulate feminist praxis for global justice.

We can read Simpson's radical resurgence with Marcella Althaus-Reid's Indecent Theology because of Althaus-Reid's emphasis on gender and sexuality in doing theology and revolutionary method to rediscover God. Indecent Theology is a radical resurgence of 'what has been excluded from theology' (Althaus-Reid 2000, p. 45) such as sexual metaphors and sexual understandings that can unfold revolutionary praxis for global justice. The revolutionary vision and radical commitment to global justice arise from divorcing the old way of doing theology but embracing 'theology as a sexual act' (p. 124). Again, Indecent Theology debunks male desires and heteropatriarchy deeply embedded in systematic theology that imagines solidarity with the poor only in terms of 'homosociability or made in His image and likeness' (p. 90). Hence, those who are considered outside God's image, such as sex workers, women of colour, and Indigenous people are excluded from the band of the poor and solidarity imagined by liberation theologians.

How would Indecent Theology move us to the radical

resurgence of the Jesus movement as a transnational feminist solidarity movement for global justice? My answer to this question is twofold. First, transnational feminist solidarity is always in the process of making with new epistemologies unfolding the concrete and materialistic knowledge of sexualized and gendered oppression, just as Indecent Theology challenges sexually neutral systematic theology. This knowledge may not reveal new realities of the poor but rather conscientize what we have not seen before.

For instance, the naked and broken body of Jesus on the cross exemplifies the sexualized death of the poor. The poor man, who challenged the relations of imperial rule (in this case, the relationship between the Roman Empire and the Indigenous in first-century Palestine), was stripped of dignity at his death. Through the sexual stories of Jesus, we can unearth the sexual and gendered realities of the poor hidden under the Covid rhetoric – 'we are in this together'. Jesus' solidarity with the poor is not merely metaphoric or spiritual but material and physical. Slowly dying on the cross, Jesus materializes the Roman Empire's necropower.

Jesus' mother, Mary, says yes to the first angel, the stranger, who appears in her room in the middle of the night (Althaus-Reid 2000). Young Mary shares a precarious life with many sex workers who say yes to strangers, again and again, to survive despite the dangerous possibility of inviting Covid-19 to their bodies.

Solidarity should not morally idealize the poor. There is nothing moral in systemic poverty and suffering. Instead, we should consciously trace survival skills and wisdom that the poor have transgenerationally and transnationally accumulated so that we can resurge and contextualize them in our own time. Sometimes in the face of extreme forms of violence, survival itself is the best resistance just as the survival and presence of Indigenous people haunt the history of European colonialism ideologically supported by Christianity.

Second, transnational feminist solidarity is messy and chaotic because sexual stories of women, sexual minorities, Indigenous peoples, and the global poor are messy and traumatized by heteropatriarchal relations of ruling, and thus, fragmented, if not compartmentalized. Our work for solidarity is not to erase messiness or unify these people for a commonly shared political goal. Instead, we navigate the fragmented stories that have survived imperial persecution. We repeatedly stitch them together to remember how structural violence has broken the bodies of the poor historically across diverse geographic locations, and to resurge and retell the stories of their resistance and survival. These stories hold pieces from Rahab, who sold her sex to support her family and finally her life; migrant woman Ruth, who used her sexuality to survive in a new land; Mary Magdalene, whose allegedly prostituted body has been glorified by the church authority for centuries; street workers in Bangkok, who make money through sexual acts to support their brothers' ordination in Buddhism; migrant sex workers around US military bases in South Korea who have children in their home countries, and so forth. We should also be aware that justice is always in the making with new practices and interpretations. Making justice is a messy business, just as it is messy to stitch fragmented and compartmentalized sexual stories together without losing their particularities or lumping them into the universalized story of misery.

As Simpson argues, a radical resurgence movement queers resurgence movements to create solidarity with other resurgence movements. What brings radical resurgence movements together is shared methodologies: critical analysis of the global power structure through the intersectionality of race, gender, class and sexuality, interrogation of the imperial operation of neoliberal capitalism, and creative retrieval of survival wisdom and struggles. These methodologies enable us to feel interconnected to one another, refuse to accept the status quo, and (re)imagine another world, deeply rooted in the survival wisdom of the poor. Through shared methodologies, specifically involved in gender and sexuality, women in similar contexts across the globe feel connected not only in experiences of structural violence but also in the struggle of liberative justice.

Indecent Theology's emphasis on the embodied experiences of the poor may create a space where messiness and chaos are valued and patiently and consciously navigated. Althaus-Reid's reflection on the Eucharist through sexual metaphors shows how to navigate the messiness of sexual stories and be united with God passionately and intimately. Although the decent church teaches the Eucharist as the space of solidarity where participants create a community of one body by sharing bread and wine, the production of bread and wine is 'hierarchical and profitable', excluding and exploiting others (Althaus-Reid, p. 92). If we see bread and wine on the communion table through the lens of a class struggle and the exploitation of the earth, the fetishism of bread and wine only reveals our desire for an imperial God. The whole ritual of the Eucharist becomes militaristic and hierarchical. A priest becomes a high commander who orders the congregation to perform precise acts and words to worship an imperial God. The militaristic, confirmative and uniformed Eucharist creates a false image of solidarity and detaches us from passionate and intimate love that God has shown to us. However, Althaus-Reid argues that just as every biblical text carries with it a subversive version, the Eucharist also bears 'intertextuality or intersexuality with God who becomes our bodies and shares our complex sexualities' (Althaus-Reid, p. 92). Here, God becomes chaos in the smell of our bodies, fluids, and hardening muscles when we make love. Through sexual metaphors, we can imagine how to feel God in unexpected times and places, passionately love God as if God were our bodies, and share God with others without harming and exploiting them or possessing God.

Sexual metaphors (and not war metaphors) can help us see practical hope in the time of a pandemic. Let us think of the policy of social distancing. The policy is necessary to stop the spread of the virus but at the same time, limits our ability to express touching others as a way of affection. Subversively, Covid challenges us to think of intimate and passionate love for our community and church differently and creatively. Love can be expressed through the affectionate gaze and genuinely caring about their well-being and resurgence movements. Wrestling with Covid, the world could see that we are more connected than we think. Witnessing the virus proportionately affecting those who are hit by systemic poverty, racism and compulsory heterosexuality, the world is awakened to the necessity of the just social system.

BLM movements across the globe, in solidarity with black Americans, are one of many examples of radical resurgence movements as global solidarity movements. Ordinary people are not quiet about systemic injustice but become more conscious of the intersectionality of different forms of systemic injustice. They see one another in their struggles for survival and global justice. A deep sense of interconnectedness is unfolded better through sexual metaphors, the erotic and aesthetic God – 'we are connected to the Divine through our connections with each other' (Alexander 2006, p. 208).

Conclusion

To interrogate inequalities and uncertainties caused by Covid-19, we should pay attention to the sexualized and gendered experiences of the global poor. Through their embodied knowledge of systemic injustice, we can see that the world leaders' war metaphors against Covid-19 are the masculinist agenda buttressed by necropower over the poor and nature. We will never win the war against Covid-19. The sexual stories of the poor may help us envision, act upon and reimagine God's radical justice that requires to overcome the hierarchical binary system and to ridicule and outsmart the system. Covid-19 manifests the current global political and economic system's failure. The decency that disguises the aggressive masculinist heteropatriarchal global political economy does not work. As Althaus-Reid (2000) eloquently revealed, to create an alternative system, we should learn to do theology grounded in indecency.

An alternative system for the post-Covid time relies on indecent people's survival wisdom whose sexuality has been demoralized, exploited, and mutilated. The traces left by these people will pave a way to the radical resurgence of Jesus' alternative system of life that has been undermined by decent systematic theology.

Note

1 Old age can be considered a form of disability in a capitalist society. From a Marxist feminist perspective, older people are marginalized because the labour market cannot or does not use them just as people with severe disabilities are not used. Those whom the labour market does not use should prove their usefulness, otherwise marginalization of these people exasperates and they are not recognized as fully human but as burdensome or surplus by capitalist society.

References

M. J. Alexander, 2006, *Pedagogies of Crossing: Meditations on Feminism, Sexual Politics, Memory, and the Sacred*, Durham: Duke University Press.
———— and C. Mohanty, 2010, 'Cartographies of Knowledge and Power: Transnational Feminism as Radical Praxis', in R. Nagar and S. Swarr (ed.), *Critical Transnational Feminist Praxis*, pp. 23–45, Albany: State University of New York Press.
Marcella Althaus-Reid, 2000, *Indecent Theology: Theological Perversions in Sex, Gender, and Politics*, London and New York: Routledge.
S. Kim, J. Kim, Y. Park, S, Kim, and C. Kim, 2020, 'Gender Analysis of Covid-19 Outbreak in South Korea: A Common Challenge and Call for Action', *Health Education and Behavior* 47.4: 525–30 (doi:10.1177/1090198120931443).
J. Lee, 2010, *Service Economies: Militarism, Sex Work, and Migrant Labor in South Korea*, Minneapolis: University of Minnesota Press.
H. Long, 2020, 'Virtual School Has Largely Forced Moms, Not Dads, to Quit Work. It Will Hurt the Economy for Years', *Washington Post* (accessed 18.12.20: www.washingtonpost.com/road-to-recovery/2020/11/06/women-workforce-jobs-report/).
Achille Mbembe, 2019, *Necropolitics*, translated by S. Corcoran, Durham: Duke University Press.
Sonya B. Norman and Shira Maguen, n.d., *Moral Injury*, US Department of Veterans Affairs, PTSD: National Center for PTSD (accessed 18.12.20: www.ptsd.va.gov/professional/treat/cooccurring/moral_injury.asp).
K. C. Pae, 2020, 'Spiritual Activism as Interfaith Dialogue: When Military Prostitution Matters', *Journal of Feminist Studies in Religion* 36.1: 71–84 (accessed 22.4.20: https://muse.jhu.edu/article/753724).
A. Rogin and A. Nawaz, 2020, '"We've Been Through This Before": Why Anti-Asian Hate Crimes are Rising amid Coronavirus' (accessed

18.12.20: www.pbs.org/newshour/nation/we-have-been-through-this-before-why-anti-asian-hate-crimes-are-rising-amid-coronavirus).

Jonathan Shay, 2014, 'Moral Injury', *Psychanalytic Psychology* 31, no.2: 182–91.

L. Simpson, 2017, *As We Have Always Done: Indigenous Freedom through Radical Resurgence*, Minneapolis: University of Minnesota Press.

R. Singer, N. Crooks and A. K. Johnson, 2020, 'Covid-19 Prevention and Protecting Sex Workers: A Call to Action', *Archives of Sexual Behavior* 49: 2739–41 (doi:10.1007/s10508-020-01849-x).

P. Watson, S. Norman, S. Maguen and J. Hamblen, 2020, *Moral Injury in Health Care Works*, US Department of Veteran Affairs, PTSD: National Center for PTSD (accessed 18.12.20: www.ptsd.va.gov/professional/treat/cooccurring/moral_injury_hcw.asp).

S. Weil, 'Gendering-Coronavirus (Covid-19) and Femicide', *The European Sociologist* 45.1. (accessed 18.12.20: www.europeansoci ologist.org/issue-45-pandemic-impossibilities-vol-1/gendering-corona virus-covid-19-and-femicide).

17

Speaking of God:
Unruly God-Talk with
Julian of Norwich

MICHAEL MAWSON

By February of 2020 it had become apparent that Covid-19 would not be contained to China and nearby countries. The world was facing a full-scale global pandemic. Many people's initial response, especially in the West, was to begin stockpiling food and other 'essential' items. Stories began to emerge about heated arguments and fights breaking out in supermarkets and department stores, often over basic foodstuffs and dwindling supplies of toilet paper (Nobel 2020). Despite the fact that there were no actual problems with either production or supply, panic-buying and stockpiling began creating shortages.

The problems caused by Covid have of course extended far beyond this consumer stockpiling. As well as its direct impact through illness and loss of life, the secondary effects of the pandemic have been catastrophic. Attempts to restrict the virus' spread through closing borders, social distancing and lockdowns have devastated national and local economies.[1] This in turn has led to higher levels of unemployment, economic hardship and financial insecurity, all of which have disproportionately affected women, black and indigenous peoples, refugees and asylum seekers and other vulnerable groups (Yiannakis 2020).

Many of our political leaders have responded in ways that once again display a scarcity mentality. To take an example, in a speech from June the Australian prime minister Scott

Morrison unveiled plans to increase Australia's investment in long-range missiles and armaments. Pointing to growing tensions and competition for resources in the Asia-Pacific region, Morrison touted such investment as necessary in order to 'prepare for a post-covid world that is poorer, more dangerous and more disorderly' (Grattan 2020). According to Morrison, Australia must fight to secure and defend its own interests in this brave new world.

In a different context, the former US president Donald Trump pressed repeatedly throughout 2020 for the reopening of schools and businesses, even as numbers of infections and deaths continued to climb. In April, taking his cue from the president, the Lt Governor of Texas, Dan Patrick, infamously insisted in an interview with Fox News that America needed to reopen, even if this would be at the expense of the lives of older and more vulnerable persons: 'Let's get back to living ... And those of us who are 70-plus, we'll take care of ourselves, but don't sacrifice the country' (Sonmez 2020). By Trump and Patrick's reasoning, the needs of the many outweigh the lives of the few. 'If that is the exchange', Patrick insists, 'I'm all in.'

A mentality of competition and scarcity is elsewhere evident in the widescale, ongoing restructuring and downsizing being embraced by businesses, educational institutions and churches. For example, in the West many universities have been radically reducing their staff, often by closing programmes and departments in areas deemed expendable. By October 2020 the pandemic had already resulted in the loss of an estimated 11,000 jobs in Australia (Karp 2020). British and American universities are anticipating and making similar levels of cuts (Lyall 2020).

In light of the impact of the Covid pandemic and the responses by politicians and other leaders, how do we as Christians, theologians and activists find a way forward? How do we talk about God in the midst of a scarcity mentality and these appeals to the need for rationalizing and defending resources?

Julian of Norwich's Context

On 13 May 1373, at the age of 30, the woman later known as Julian of Norwich received a series of 16 dense, overlapping visions, *'shewings'*, centred around the Passion of Christ as the place of God's abundant love. She received her visions while suffering an unknown illness over several days, while believing herself to be on her deathbed. Upon recovering, she recorded her visions in a short document, *A Vision Shown to a Devout Woman*, now usually referred to as Julian's 'Short Text'. Over subsequent decades, living in relative social isolation as an anchoress at the Church of St Julian in Norwich, she composed a series of longer meditations on her visions and their possible meanings: *A Revelation of Divine Love*, referred to as her 'Long Text'.[2]

There are some striking parallels between Julian's context and our Covid context. Julian received her visions and composed her reflections in a world being remade following a pandemic. The black death, or bubonic plague, had ravaged Europe between 1347 and 1351, when she was a child. Estimates are that more than a third of the population of Norwich died as a direct result of this plague; 'people died, horribly and suddenly and in great numbers' (Jantzen 1987, p. 8). As Jantzen continues, the trauma and 'psychological impact on survivors was incalculable, made worse in subsequent years by the further outbreaks which occurred at unpredictable intervals' (Jantzen, p. 7). Many of those who had died had not been able to receive their final rites, placing in question their very salvation.

Julian's context was also marked by political protest and revolution, notably the Peasants' Revolt of 1381. 'She received her first visions', Amy Laura Hall notes, 'during a time when peasants, calling for an end to serfdom, burned buildings and even invaded the Tower of London' (Hall 2012, p. 158). The instability and desperation caused by plague and famine had created a situation in which some were willing to risk open rebellion. This rebellion extended to Norwich itself. In June 1381, Geoffrey Litster led a small group of rebels to capture and briefly occupy Norwich Castle. Litster and his comrades were quickly rounded up, put on trial and executed by the Bishop of Norwich, Henry le Despenser (Jantzen, pp. 8–9).

As in our own Covid context, political leaders sought to use both plague and rebellion as cover for accumulating and extending power.[3] This wider period was marked by an increased centralization of power in the hands of a small number of monarchies and the papacy. This process is reflected in the series of conflicts now known as the Hundred Years' War, waged between and within England and France. This war broke out in 1337, six years before Julian's birth, and finally concluded several decades after her death. Continual warfare required both a constant supply of soldiers for armies and funds in the form of taxation, placing further pressure on already decimated and desperate local populations (Jantzen, p. 10).

All of these form the broader background of Julian's life and theology. She received her visions and composed her meditations in the midst of trauma and these challenges. Even while making little direct reference to this context, the substance and forms of Julian's theology were contesting many of the basic assumptions of those holding and exerting power. This is especially the case with respect to her central insistence on the sheer abundance of God's love and grace.

Julian's Theology and Language of Abundance

Julian's theology is a theology of God's goodness and abundance. Her visions and meditations consistently witness to a God who intimately embraces and lovingly upholds humanity: 'Our Lord showed me ... how intimately he loves us. I saw that he is everything that is good and comforting and helpful to us. He is our clothing that enwraps us and enfolds us, embraces us and wholly encloses us, surrounding us out of tender love, so that he can never leave us' (Julian 2015, p. 45). For Julian, God's love reaches down and tenderly embraces all human beings, without regard for worldly boundaries or hierarchies.

Julian's visions of this intimate, boundary-crossing love focus upon the cross. She received her visions while lying in her bed gazing at a crucifix which a parson held before her (Julian, p. 5). And most of her visions focused upon Christ's bodily suffering and dying. Indeed, in one of her most memorable visions, she sees copious amounts of blood pouring forth from Christ's wounds following his scourging: 'The hot blood ran so abundantly that neither skin nor wound was to be seen … And it was so abundant to my way of seeing that it seemed to be that if it had been the real thing in nature and essence … it would have saturated the bed with blood and overflowed all around' (Julian, p. 56). What is the significance of this overflowing, saturating blood? In a context where bloodlines were carefully regulated and where access to the Eucharist was highly restricted, Julian has a vision of Christ's blood being continually poured out and made available to all (see Hall 2018, pp. 61–79).

If the substance or content of many of Julian's visions witnesses to God's abundant love in Christ, the very forms of her theology similarly testify to this abundance. Julian's dense, complex language and imagery continually exceeds and disrupts more linear and conventional ways of speaking about God and the world.[4] Put differently, she refuses to restrain herself to more accepted and acceptable ways of doing theology. Here it is possible to draw attention to only a few of the abundant and unruly features of her theological language.

First, it is significant that Julian composed her theology in the vernacular, in English. (Julian has often been credited as the first woman to have written a book in English.) Rather than using the more respected and precise forms of Latin, she opted for language that was riskier, messier, closer to the ground: 'Vernacular theology describes Christian writings by people without explicit churchly power, written in the language that people working in fields would speak, if not read' (Hall 2018, p. 42). Julian sought to communicate God's abundant love using the common tongue, with all its hybridities, inconsistencies and neologisms.

In addition, Julian uses language in ways that are often playful and poetic. Reading Julian as a preacher and performance artist, Donyelle McCray has recently drawn attention to some stylistic features of Julian's writing, to the ways she uses 'rhyme, meter, and alliteration' (McCray 2019, p. 28). As an example, McCray points to Julian's use of alliteration and rhyme in her descriptions of Jesus' dying in her eighth vision: 'And at the beginning, while the flesh was fresh and bleeding, the continual pressure of the thorns made the wounds wide' (Julian, p. 63). McCray suggests that Julian's language is performative; rather than simply representing or describing God's love, 'Julian seeks to create an experience that is both cognitive and affective, one that will lead to spiritual formation and ethical development' (McCray, p. 28).[5]

There is linguistic abundance in how Julian folds and collapses her images into one another. In a famous reflection on her fourteenth vision, she describes Christ as a mother who 'gives birth to us into joy and to endless life'. Mixing her metaphors, Julian proceeds to suggest that as a mother Christ 'can lead us intimately into his blessed breast through his sweet open side and reveal within part of the Godhead and the joys of heaven' (Julian, pp. 130–1). Julian's maternal image is here enfolded into Christ's wounds, transgressing gender norms and destabilizing bodily and linguistic boundaries.[6] Julian layers and enfolds images in ways that are bewildering and disorientating for the reader.

Moreover, Julian is at times unwilling to confine herself to a single line of interpretation. She instead attributes different, even conflicting meanings to some of her visions. One place this is apparent is her lengthy reflection on the parable of the Lord and the servant, again in her fourteenth vision. Having provided a first reading of this parable, in which she identifies the servant as Adam, Julian reflects: 'I could not understand it fully and to my satisfaction; for in the servant who represented Adam, I saw … characteristics which could in no way be attributed to Adam alone' (Julian, p. 108). Accordingly, she provides a second, entirely different reading, in which she identifies the servant as Christ. Julian juxtaposes and layers these two readings without attempting to reconcile them. The sheer abundance of meaning within the parable cannot be contained to one interpretation.[7]

Finally, Julian employs apocalyptic language in her theology. She consciously makes use of the kinds of abundant. excessive polemics and imagery found in Daniel 7 and the book of Revelation.[8] Reflecting on her fifth vision, she writes that the devil will be 'scorned at the Last Judgement by al. those who will be saved, whose consolation he greatly envies … All the misery and tribulation that he would have liked tc bring upon them shall go with him forever to hell' (Julian. p. 59).[9] She elsewhere imagines the devil as continually tempting her to despair of God's love and grace. With the devil's final judgement, all temptation to despair shall be overcome. Like John on Patmos, in the midst of persecution and suffering Julian boldly proclaims a time when 'all shall be well' (Julian, p. 22) or when 'he will wipe every tear from their eyes' (Rev. 2.14).

The Politics of Unruly God-Talk

There is a politics inherent in Julian's abundant, unruly language. Simply by speaking of God in these complex and excessive ways Julian contests and displaces many of the assumptions of those aiming to wield power and enforce order. Julian's language of abundance contests language of scarcity and competition, as well as the assumption that violence and control, rather than love, is the necessary response to trauma and a disorderly world.

To be clear, Julian does not directly challenge the ecclesial and political authorities of her day. Indeed, she is often rather quick to proclaim her own orthodoxy and willingness to submit: 'In all things I believe as Holy Church teaches' (Julian, p. 10).[10] Julian's form of resistance is subtler and closer to the ground. She simply writes in unruly ways of God's love in Christ; but in so doing, she breaks with and exceeds the kinds of linear language and thinking that facilitates and regulates control from above.

More recently, Judith Butler has drawn attention to how linear, common-sense language and thinking on some level always serves the status quo. In early 1999, the conservative journal *Philosophy and Literature* had named Butler as the annual winner of their 'Bad Writer' competition. This journal had invited its readers to submit the 'ugliest most stylistically awful' sentences that they had discovered during the preceding year. This naming and shaming of Butler as a 'bad writer' received widespread coverage in the United States, both in academic and more mainstream newspapers.

In March 1999, Butler responded with an op-ed in the *New York Times*: 'A "Bad Writer" Bites Back' (Butler 1999). She first notes that the various targets of *Philosophy and Literature* in preceding years had invariably been 'scholars on the left whose work focuses on topics like sexuality, race, nationalism and the workings of capitalism'. In other words, these accusations of 'bad writing' were not simply politically innocent or neutral. This leads Butler to reflect, 'Why are some of the most trenchant social criticisms often expressed through difficult and demanding language?'

In her op-ed, Butler further suggests that by its very nature critical scholarship must involve reflecting upon and interrogating conventional and commonly held beliefs: 'scholars are obliged to question common sense, interrogate its tacit presumptions and provoke new ways of looking at a familiar world'. This is because much of what passes for common sense is itself highly questionable: 'For decades of American history, it was "common sense" in some quarters for white people to own slaves and for women not to vote' (Butler 1999). Appeals to common sense and straightforward language, therefore, too easily facilitate injustice by preventing critical thinking and preserving the status quo.

Further, Butler insists that interrogating common sense assumptions will necessarily require new and unfamiliar patterns of thinking and writing. And this in turn will require the kinds of language that often break with and fall outside what is considered conventional and commonsensical. On this point she quotes Herbert Marcuse: 'if what he [the social critic] says could be said in terms of ordinary language he would probably have done so in the first place' (Butler 1999). In other words, imagining and articulating how things might begin to be otherwise will require new, sometimes unruly forms of writing.

Accordingly, Butler can help us to recognize what is at stake with Julian's excessive, unruly theological language. Julian's unruly God-talk is itself central for how she imagines and proclaims God's abundant love in the midst of trauma and violence. Julian's vernacular theology – with its playful and poetic style, layering and enfolding of images, multiple lines of interpretation and apocalypticism – is integral to how she imagines and proclaims a new and different reality, one in which people are joined together in God's love and Christ's blood.[11] In other words, Julian's very unwillingness to follow the rules that constitute 'good' theology and writing is itself a form of protest.

Conclusion

What can we learn from Julian's excessive, unruly ways of talking of God? How can her abundant language assist us in the midst of Covid and with responding to the many challenges that we are facing today? Julian's politics of linguistic abundance, I suggest, can continue to engender and support rich, complex forms of resistance and protest. More specifically, attending to Julian's unruly language and theology can assist with unravelling and overcoming assumptions of scarcity and competition in our own contexts,[12] as well as appeals to the need for self-preservation through either stockpiling or defending resources.

In the context of the Covid crisis, some of us will no doubt have options for direct resistance and opposition.[13] Those of us with a relative level of privilege will have options for contesting language of scarcity and rationalization outright, as well as the ways in which our politicians and other leaders are drawing upon and mobilizing such language. We will be in positions to invoke norms of justice, rights and equality in order to push back against this kind of rhetoric and these troubling developments. Or else we will be in positions to organize and join demonstrations in direct solidarity with those who are experiencing actual scarcity and oppression.[14]

For other people, however, more direct forms of protest and resistance may not be so readily available and viable. For those living under oppressive regimes, or in more precarious situations, resistance and protest will at times need to proceed in ways that are messier and closer to the ground. And this may in turn involve adopting and deploying excessive and abundant language of precisely the kind displayed by Julian. Attending to Julian's unruly God-talk, therefore, may help with both recognizing and cultivating some of these messier, more complex forms of resistance and protest.

For all of us, Julian shows how speaking of an abundant God – and how speaking abundantly of God – can itself challenge and displace the logics and language of scarcity and competition. She reminds us that we worship a God whose love is continually transgressing human boundaries and hierarchies. Julian's unruly language and theology witnesses to an abundance that exceeds and scrambles all attempts at economic rationalization and political control.[15] Through God's abundant love in Christ, all human beings are joined together and affirmed as worthy of love and protection.

As part of Julian's very last vision she receives a promise and assurance from God: 'You shall not be overcome.' These words, she reflects, 'were said very distinctly and very powerfully against the tribulations that may come. He did not say, "you shall not be perturbed, you shall not be troubled, you shall not be distressed", but he said, "you shall not be overcome"' (Julian, p. 143). As we respond to our own contexts and the many challenges that we are facing, may we too hold in faith that we shall not be overcome.

Notes

1 On the economic impact of Covid, see Chapter 7 in this volume, by Gerald O. West. West focuses on the context of South Africa.

2 In this chapter I move back and forth between Julian's short text

and her long text. I use Barry Windeatt's translation for all quotations from Julian, while also consulting the old English in the Jenkins and Watson edition.

3 Examples in our own context include Hong Kong, Thailand, Algeria, Hungary and Israel-Palestine (McLaughlin 2020).

4 The obvious point of comparison in Julian's context would be with the logical forms and linguistic precision of the scholastic theology of someone like Thomas Aquinas.

5 See Judith Butler on the performativity of language (Butler 1997, pp. 1–41). Note that McCray briefly makes a link to Butler when drawing out the bodily nature of speech and writing (McCray, p. 60).

6 Here we see resonance with Marcella Althaus-Reid's more recent 'indecent theology', which similarly aims to contest and transgress accepted and acceptable norms and boundaries in and beyond theology (Althaus-Reid 2000). On Althaus-Reid's theology, see also Chapter 16, by Christine Pae.

7 See Denys Turner's rich exposition of Julian's reflections on this parable (Turner 2011, p. 103–34).

8 In a recent study Justin Byron-Davies argued that Julian is drawing extensively on the biblical book of Revelation in her theology (Byron-Davies 2020).

9 It is worth noting that in contrast to John of Patmos, however, Julian largely reserves hell and the final judgement for the devil and sin.

10 Does Julian proclaim her obedience just a little too loudly and too quickly? Insights from postcolonial theory into 'mimicry' might be useful for reading Julian's proclamations of orthodoxy and submission (see Bhabha 1994, p. 90).

11 Hall draws out the kind of solidarity implied by Julian's theology. Her visions invite 'people, in spite of all of the risks, to see one another as kin, in the foolish safety of God's present. Her visions address domination from the vantage of someone who has been underneath the structures that divide people from each other and from within the hope that change is coming' (Hall 2018, p. 50).

12 My interest in this chapter is in how appeals to scarcity are being made in Western and more affluent contexts. That is, in how such appeals are used to justify a range of measures from consumer stockpiling through to increasing armaments and security. This is not intended to in any way deny or downplay situations of actual scarcity and hardship, either within these Western countries or in other parts of the world.

13 Even these direct forms of resistance can be supported and deepened by unruly language and theology like that of Julian. There need not be a choice or clear line between direct and indirect forms of resistance and protest.

14 See, for example, Chapter 4, by Angelica Tostes and Delana Corazza.

15 Amy Laura Hall has brilliantly drawn out how Julian's vision of God 'scrambles calculation and the usual ways we weigh, measure and sort ourselves, strangers, our children, our churches, or our neighbourhoods (Hall 2018, p.27).

References

Marcella Althaus-Reid, 2000, *Indecent Theology*, London: Routledge.

Homi Bhabha, 1994, *The Location of Culture*, London: Routledge.

Judith Butler, 1997, *Excitable Speech: A Politics of the Performative*, London: Routledge.

———, 1999, 'A Bad Writer Bites Back', *New York Times*, 20 March 20 (accessed 7.11.20: https://archive.nytimes.com/query.nytimes.com/gst/fullpage-950CE5D61531F933A15750C0A96F958260.html).

Justin Byron-Davies, 2020, *Revelation and the Apocalypse in Late Medieval Literature: Writings of Julian of Norwich and William Langland*, Cardiff: University of Wales Press.

Michelle Grattan, 2020, 'Scott Morrison pivots Australian Defence Force to meet more threatening regional outlook', *The Conversation*, 30 June (accessed 7.11.20: https://theconversation.com/scott-morrison-pivots-australian-defence-force-to-meet-more-threatening-regional-outlook-141727).

Amy Laura Hall, 2012, 'A Ravishing and Restful Sight: Seeing with Julian of Norwich', in Brian Brock and John Swinton (eds), *Disability in the Christian Tradition: A Reader*, pp. 152–83, Grand Rapids, MI: Eerdmans.

———, 2018, *Laughing at the Devil: Seeing the World with Julian of Norwich*, Durham, NC: Duke University Press.

Grace Jantzen, 1987, *Julian of Norwich: Mystic and Theologian*, London: SPCK.

Jacqueline Jenkins (ed.), 2007, *The Writings of Julian of Norwich: A Vision Showed to a Devout Woman and A Revelation of Love*, translated by Nicholas Watson, Penn State University Press.

Julian of Norwich, 2015, *Revelations of Divine Love*, translated by Barry Windeatt, Oxford: Oxford University Press.

Paul Karp, 2020, 'Almost 500 More University Jobs to go at ANU and UNSW as Covid Cuts Bite', *The Guardian*, 16 September (accessed 7.11.20: www.theguardian.com/australia-news/2020/sep/16/almost-500-more-university-jobs-anu-unsw-covid-cuts-bite).

Sarah Lyall, 2020, 'University in Britain Brace for Cuts in Subsidies', *The New York Times*, 15 October (accessed 7.11.20: www.nytimes.com/2010/10/16/world/europe/16britain.html).

Donyelle C. McCray, 2019, *The Censored Pulpit: Julian of Norwich as Preacher*, Lanham, MD: Lexington/ Fortress Academic.

Timothy McLaughlin, 2020, 'Where the Virus is Cover for Authoritarianism', *The Atlantic*, 25 August (accessed 7.11.20: www.theatlantic.com/international/archive/2020/08/pandemic-protest-double-standard-authoritarianism/615622/).

Emma Nobel, 2020, 'Coronavirus toilet paper stockpiling frenzy ignores our long history without the soft stuff', *ABC News*, 15 March (accessed 7.11.20: www.abc.net.au/news/2020-03-15/coronavirus-toilet-paper-on-sale-covid-19-unlikely-history/12054636).

Felicia Sonmez, 2020, 'Texas Lt. Gov. Dan Patrick comes under fire', *The Washington Post*, 25 March (accessed 7.11.20: www.washingtonpost.com/politics/texas-lt-gov-dan-patrick-comes-under-fire-for-saying-seniors-should-take-a-chance-on-their-own-lives-for-sake-of-grandchildren-during-coronavirus-crisis/2020/03/24/e6f64858-6de6-11ea-b148-e4ce3fbd85b5_story.html).

Denys Turner, 2011, *Julian of Norwich, Theologian*, New Haven, CT: Yale University Press.

Michael Yiannakis, 2020, 'Raw Deal for Workplace Minorities Amplified by Covid-19', *The Lighthouse*, 2 November (accessed 7.11.20: https://lighthouse.mq.edu.au/article/october-2020/Raw-deal-for-workplace-minorities-amplified-by-Covid-19).

18

Lagimālie: Covid, De-Onefication of Theologies, and Eco-Relational Well-being

UPOLU LUMĀ VAAI

The call to reinvigorate eco-relational Pasifika indigenous values and ways of knowing and being since the 1960s has regained momentum after the dominant old colonial pathological *narrative of onefication* (Vaai 2020) collapsed during the Covid pandemic. Pasifika churches have been at the forefront of this *relational turn* since the 1970s through the theologies produced from the Pacific Theological College. The recent decision by Pasifika churches (Pacific Church Leaders 2017) to shift from the *unity in Christ* narrative that prioritizes church and human unity which dominated ecumenism in the region to the *household of God* (henceforth called the *cosmic Aiga*[1]) is widening the scope of ecumenism to include ecologies and economies of life that form the basis of the cosmic Aiga, as well as calls for the *de-onefication* of Pasifika theologies in order to develop a solid theological foundation for this vision.

One of the fundamental principles that holds together the cosmic Aiga is *lagimālie*, an *other-informed consciousness*, where everything cosmic is considered family and from this foundation the cosmological mindfulness and balance is upheld. *Lagimālie* is a serious consideration of the balance of the multiplicity of relationalities that constitute the whole. There is no whole if the parts are not relationally balanced and respected. This chapter aims to address the dominance of the old pathological *narrative of onefication* that has infiltrated the

Pasifika cosmic Aiga and its theological discourses. It calls for revisiting the *lagimālie consciousness* in Pasifika theologizing in situations of crisis.

Lagimālie

In the Pacific *itulagi* (side of the horizon; see Vaai and Casimira, pp. 6–7), the term *lagimālie* is made up of *lagi*, meaning 'life', and *mālie*, meaning 'wellness' or 'balance'. Hence *lagimālie* means wellness or balance of life. This balance is achieved only when there is a cosmic harmony of the self, the others, the environment, and the whole cosmos to ensure the stability of the whole (Tui Atua, pp. 137–51). It is a wrestling to understand the individual as part of the cosmic-community and the cosmic-community as imaged in the individual. *Lagimālie* is a rhythmic intuition to realize that everything is integrated, inter- and intra-related. It is a harmonious, not a monolithic, cross-flowing rhythm of cosmos and person, ecology and economy, unity and distinction, connection and difference. Some Pasifika theologians have used the *moana* metaphor to illustrate this interconnected rhythm of life that is never static nor human-centric (Halapua 2010). The entire cosmic Aiga is a body of integrated memory. This interconnected reality determines how each member of the cosmic Aiga responds to any imbalance in the body, for example the imbalance caused by pandemics.

This *holistic gaze* has however been disrupted by the *narrative of onefication*, a narrative that has introduced a non-holistic way of approaching life, especially the focus on splitting into compartmentalization which we now know does not work during disasters such as pandemics. Today we are so obsessed with categorizing people, cultures, knowledge and God into fixed systems. This *onefication* agenda in theology has for many years encouraged a view of life through split categories to achieve one answer, one truth, or one destination. This single-strandic theologizing has pushed us to define and prove whether God systematically fits into split categories or strands such as a priori or posteriori, objective or subjective, substance or relation, male or female, process or solitary, heaven or earth. This has created incompatibilities between our Christian theology of God and our search for answers during crises. The systematic God we have created seems undeterred by the cries of those affected by the pandemic.

In Christology we are urged to prove whether Christ systematically fits into categories of either divinity or humanity, centre or margin, history or faith. In ecumenism, we promote an unhealthy separation of the *oikos triplets* of economy, ecology and oikoumene. Economy is assumed to be owned by the neo-liberal capitalist empire who turned it into a money-making machine, ecology by the scientific empire who turned it into a mere object to be studied, and oikoumene by the Christian empire who turned it into a Christian tradition to serve human and church unity. In ecumenical history we assume that oikoumene has nothing to do with economic and ecological dilemmas. How did we end up in this compartmentalized mentality?

Today, we have many Pasifika reductionist theologies that choose powerful categories such as divine, centre, male, objective, pure, solitary and normality, which are dysfunctional during pandemics. Those choices disallow us to recognize the ambiguity in God where the pleasant can be found in the unpleasant, the possibility in the impossibility, the comfort in the bizarre, and the more in the less. Theologies shaped by split categories have packaged up God into powerful rigid systematic explanations that deny God's active and dynamic flow and as a result prevented those who suffer from illnesses to faithfully engage with God, even doubting and questioning God.

We can see this *onefication* approach in Pasifika economies as well especially in those who depend on tourism. Many countries are struggling since borders were closed and international movements put on hold. Some governments such as Papua New Guinea, Samoa and Fiji have used the pandemic crisis to regain confidence in a government in crisis. Some have used the normal exclusivist political approach to fend off any critique of mishandling the pandemic. Some governments such as

Fiji have gone to the extent of passing laws that give employers the power to lay off or terminate an employee during the pandemic period. While the pandemic is deeply affecting every household, those in power are weaponizing the pandemic to secure investment.

In pushing to centralize resources and benefits, policies are favoured over life, entitlements over responsibilities, and political and economic systems over communities and people. We have been pushing the narrative: the survival of the many lives is dependent upon the survival of a human-centric epoch, the so-called Anthropocene, or the capitalist cultural epoch, the Capitalocene. So when pandemics happen, the cosmic Aiga is in a state of cosmic imbalance. Continual discussions of whether God has contributed to this imbalance is critical in order to strengthen our vision of God who is both distant and close. This ambiguity in God will always enrich our discussion of divine providence. In *lagimālie*, unless the cosmic Aiga restores its eco-relational balance, pandemics will become normal and we head towards self-destruction.

While this narrative has its own place in the dominant theological tradition, it is ill-equipped, at least from a Pasifika eco-relational perspective, to diagnose theological and social problems, much less to offer alternatives of well-being during extreme circumstances such as pandemics. Compartmentalization removes God to the realm of lordship and objective logic. As a result, it warrants an either/or way of thinking that favours the autonomous self over cosmic relationships, linearity over complexity, hegemony over multiplicity and diversity. So when pandemics or natural disasters strike, communal safety nets are scarce and the individual is presented with very limited options for survival and for building resilience.

There are three *lagimālie* imperatives I want to highlight that could assist in restoring the eco-relational balance of the cosmic Aiga.

Everything is Flesh, Bones and Blood

First is the *whole of life* imperative. It approaches life from a holistic perspective, and requires radical reconstruction and *de-onefication* of Pasifika theologies. In this perspective, God is relational and the story of the cosmos must be primary. The redemption story is therefore subordinate to the creation story (Wallace 2019). To be precise, creation (focus on all of life) precedes redemption (focus on redeeming human life). Cosmology precedes anthropology. Relations precede being. Before there was grace to save human beings, there was prevenient grace that organized and animated the communion of creation. And because of this cosmic grace that encompasses all, everything is genealogically connected through the Spirit. In this light, our well-being is made possible by cosmic well-being. We cannot talk about well-being of the parts until we take seriously the well-being of the whole.

It is for this reason that Christianity should challenge the century-old anthropocentric pride that has deep links to any disaster, including the Covid pandemic. God's love should never be confined to only one member of the cosmic Aiga, the human. Because Pasifika people see everything as family – *flesh, bones and blood* – we are products of this eco-relational *cosmic genealogical connection* where one life contains the dimension of everything else. Humans for example have land and ocean dimensions. Maori people see this whakapapa (or lineage) as a tool within culture that traces the relationship between all of creation (Cadigan 2010). This is why most Pasifika terms for earth – such as *eleele, vanua, whenua, palapala, aba, 'āina* – are synonyms for personhood. I therefore speak as the Earth; I don't speak for the Earth. Everyday economic activities – whether it's producing food, fetching water, healing, or ceremonial events – are informed by this cosmic connection. Thus the question is not so much about whether God is part of us as we already exist 'in' God, but rather about how we exist and struggle to understand the integrated multiple eco-relationships that constitute the whole of life 'in' God.

Because our eco-theologies are unable to push cosmology as

primary, we fail to account for the *cosmic genealogical connection* of life. Many eco-theologians argue that we should redeem ecology from this *onefication* narrative that compartmentalizes life in order to consider the harmony of the whole. However, it is not enough, as many Pasifika eco-theologians argue, to have an integrated model of ecology that pushes for the centrality of human relationship with creation until we recognize this *cosmic genealogical connection* that we hold with everything else (see Boseto 1995a and 1995b; Tofaeono 2000; Tuwere 2002, pp. 35f.; Bird 2008).

Unless we reconstruct redemption theology to be cosmic oriented, we will be stuck in the *onefication* agenda. Because of the eco-relational structure of life, we cannot treat topics such as *ecology, economy* or *oikoumene* on their own. They are deeply interwoven touching each other. *Lagimālie* clusters around notions of complexity, multiplicity and negotiability.

Lagimālie is achieved because life is never fixed to one direction. We've seen that those who resist the fluid flow of life struggle to navigate the Covid challenges. Pandemics become worse when we treat life single-strandically as a fixed entity. *Lagimālie* finds its rhythm in the embrace of the dynamic multifarious nature of life where well-being cannot be confined to one aspect of life.

In this regard, *lagimālie* does not subscribe to the words *normal* or *new normal*. This is because life is perceived in flow and motion. Life is a dynamic movement. This resonates well with the Samoan wisdom saying, 'the wisdom of the wise is negotiable and the wisdom of the fool is fixed'. It also resonates well with the perspective of liquidity where Pasifika does not refer to 'islands in the sea' as explorers had assumed, but rather 'our sea of islands' connected and linked by the liquid flow of the blue moana, our Pasifika ocean (Hau'ofa 1993). In this respect, life is perceived as a liquid flow.

Islanders knew that there would be disasters on the journey. We live in the ambiguous space of life and death, attuned to the *let be* rather than the *must be* way of thinking where the *not yet* is accounted in the here and now. It is like thinking in the Spirit, as Leonardo Boff urges, where our constant becoming is integral to such thinking (Boff 2015, p. viii). This is why life is subject to refinement, redefinitions and new complexes of meaning or relationship. Without this 'open horizon', as Raimon Panikkar puts it (1993, p. 5), we would struggle to find life and meaning in disasters yet to come.

Seeing everything as family promotes a different methodological starting point. Instead of starting with the *human–earth–cosmos* sequence where the human being is centred, we are invited to shift to the *cosmos–earth–human* sequence; the latter is a collective eco-relational imperative to make sense of our place in the cosmic Aiga. If cosmology is a relationship and not just about systems of chromosomes and sets of natural rules, then eco-relationality calls for a move beyond mere interactions and correlations to recognizing everything as family. It calls for an eco-relational theology that seriously takes the cosmos as a genealogical relationship, a divided undivided whole where God is an integral part of that whole. In this divided undivided whole, when we refer to one member, we refer to the whole Aiga; the face of the whole is manifested in the face of the one.

This *holistic gaze* is something that the Cappadocian fathers taught us in their Trinitarian theologies, especially Gregory of Nazianzus (1978, p. 375). When we speak of the one, we speak of the whole. Because one is mutually included in the whole, when one suffers the whole also suffers. Eco-relationality critiques ecological destruction, along with the systems of relationships that contribute to the collapse and destruction of this eco-relational connections. It emphasizes the shift from space to *deep living connections* within cosmology, embracing the multi-dimensional relationships and their inherent values together with spirituality that constitutes space and time. It promotes a deeply fluid connectedness where cosmology is seen as a dimension of everything else.

We Are Therefore We Live

Second is the *we are* imperative – *tatou tatou* (Te Paa Daniel, p. 17). To say that cosmology precedes anthropology paves the way to the idea that we are meant to live according the cosmic rhythms. Because we are part of an eco-relational whole where we exist because of our deep relationship with each other, we cannot simply exist on our own. *We are therefore we live* (Vaai 2019) is the general Pasifika cosmic axiom that works during extreme circumstances such as pandemics. The focus on living collaboratively and relationally is at the heart of *lagimālie*. This urges us to shift the questions and discussions from *how God lives* into *how we live*. In *lagimālie*, instead of pushing the question of 'where is God in the pandemic' or 'why did God punish us with this pandemic', we start with a 'we are' question such as: what is our responsibility to each other during this pandemic? The 'we are' puts emphasis on community innovation and values of care.

During the pandemic for example, many families in rural areas had to fetch water from far away wells and rivers. So it's not a matter of when, why, and how to wash their hands as per Covid measures; the critical question is from where to get water. Survival and resilience under these situations of crisis are dependent upon the assistance of other people, and also from creative skills acquired from living closely to the Earth. In this case, we need to shift from the old pathological theology of *God the answer* to *God as presence*. The former forces our agenda on God where God is locked into a contractual agreement to provide the answers for our issues. The latter allows space for our faith to grow into relationships by identifying the presence of God in situations of crisis in the present. *Lagimālie* is when we know how to present God to others through our presence so that the silence of God's presence is active in our very presence.

Rather than assessing God in the Covid pandemic, we should be drawn into reassessing ourselves and our compassionate responsibilities towards the restoration and healing of others. These 'we are' actions of 'deep solidarity' (Rieger 2018) can become traces of God's presence in the world, especially at the time of vulnerability. While we must constantly analyse the deep-seated notions of God we so often uphold and defend, we should refrain from an assessment of God that may result in propagating divine judgements and unnecessary presuppositions that either split God into binary categories or frame others as recipients of divine judgement. The 'we are' approach problematizes this rationalistic direction of the theology of God and also the pietistic heavenization that dominates Pasifika Christianity where everything, including pandemics, is transferred to God's heavenly juridical abode. While theology aims at articulating divine involvement in our suffering, the 'we are' approach taken up by most grassroots communities simply offers a more practical way than that of a heaven-bound one. It draws us to focus more on the multiple skills and resilient ways as well as cheaper options that we already have within our grasp (instead of what we do not have) to heal others. Faith in this respect is more relational and practical than abstractive.

A Straight Line is Only a Curve

Third is the *relational time* imperative. This can be seen in the Pasifika processes of seeking wisdom. Achieving *lagimālie* involves a serious consideration of the concept of time, especially a lengthy and time-consuming process of consulting the complex eco-relational genealogies. Slowness in process is critical. A slow process ensures minimum consequences on life as a whole. This slow consultation is with other fellow human beings, the environment, the cosmos and the self. Asking critical questions in this process is important in order to achieve the wisdom of *lagimālie*. The Covid pandemic has taught us the importance of slowing down through 'curved time'.

Tevita Mohenoa Puloka from Tonga presented 'a straight line is a curve' axiom, arguing that the ways that Tongan and Pasifika communities think and do things go against linearity. Using the metaphor of the clothesline, 'when one unit of laundry hangs on the clothesline the line is fairly straight but when 50 units hang on the line, it curves' (Puloka 2007, p. 61). The straight line represents those who rush to push life to the future by destroying the present. The curve represents a relational multi-strandic approach to life and well-being, one that is not confined to systems and categories. It sees life through the lens of the communal unit comprised of the 'We'. When we look after each other and the whole cosmos, we are actually living in 'curved time' rather than in a straight 'linear time'. The latter normally rushes to reach conclusions. The former takes time because it is also concerned with others along the way. For this reason, any decision making will have to take a curved approach which involves a long process of consultation that includes the trees, land, ocean, people and ancestors.

In Samoa for example, relational time or curved time is uttered in the process called *moe le toa* (let the *toa* sleep) where deliberation would be postponed to the next day when a solid decision is not reached. The *toa* is a tree used for creating durable traditional weapons. For the sake of durability, it needs to sleep in water in a lengthy period of time to ensure it is tough to achieve its purpose. The more it sleeps the more it is durable. The metaphor is used for decision making in the sense that the more time a decision is challenged with questions in consultation with the earth and the ecological community the more durable it is; a *toa* decision carefully navigates through complexity to achieve a comprehensive and harmonious whole. The metaphor is also used to challenge those who insist on living a fast pace of life at the expense of everything else. Rethinking this life during and after Covid can be seen not only as the Kairos moment for the region, but also a moment to create Kairos movements to change our story (Bhagwan 2020).

Today we are inundated with linear time shaped by rapidification, according to Pope Francis (2015). While ecology moves in a slow pace, we tend to rush everything including our own decisions without careful consideration of their consequences. We are so future-oriented that more and more we push our developments to invest in the future at the expense of the present. We cut down hub trees such as those in the Melanesian forests. We are now moving towards mining the ocean as in

Papua New Guinea, Cook Islands and Tonga. The well-being of vulnerable communities is at risk.

The Covid pandemic has challenged future-oriented rapidification attitudes, urging us to slow down for the sake of the cosmos. By slowing down, we may properly assist others walking alongside us. This consideration of others can be achieved through ongoing questioning, consultative discernment and sense-making, and critical dialogue with all eco-relationships of the cosmic Aiga. Willingness to question ourselves and to consult with a discerning and listening resolve is key to the *lagimālie* of eco-relationships. The goal is not to achieve an absolute answer, but rather to institute a check and balance mechanism whereby holistic life is embraced and upheld with mutual respect and dignity. The *lagimālie* of the whole during and after the pandemic depends on this *de-onefication* wisdom process of Pasifika.

Conclusion

Situations controlled by the *narrative of onefication* suppress *lagimālie*. We thus need to move away from this normal narrative. We will do this through the *de-onefication* of our theologies and life. Focusing on one culture, one way, one truth, or one idea is unsustainable and unhealthy. What the pandemic has revealed is that this narrative has contributed to the imbalance in the cosmic Aiga.

Lagimālie is a consciousness that focuses on the *de-onefication* of life. It embraces the multiplicity of relationalities. The three imperatives ('whole of life', 'we are' and 'relational time') are grounded on the primacy of the cosmos. *Lagimālie* finds its rhythm in the embrace of the dynamic multifarious nature of life according to which well-being cannot be confined to one aspect of life. The challenge is harnessing the ability to live as a family within the scope and criteria of the *cosmic genealogical connection*. It requires us to think of what is already available within our reach to build resilience and solidarity during disasters and crises. It also requires us to critically reflect on how we understand and theologize within such crises for the sake of well-being.

Note

1 Aiga in the narrow sense refers to a social unit in the village comprised of immediate family members and kinship. In the broader sense it is a holistic Pasifika concept that refers to the cosmic extended family inclusive of land, ocean, peoples, ancestors and spirits/gods that constitute wholeness of life.

References

James Bhagwan, 2020, 'Back to the Future: Reappropriating Island Time and a Return to Kairos', in James Bhagwan *et al.* (eds), *From the Deep: Pasifiki Voices for a New Story*, pp. 37–48, Suva: Pacific Theological College.

Cliff Bird, 2008, 'Pepesa – The Household of Life: A Theological Exploration of Land in the Context of Change in Solomon Islands', PhD diss., Charles Sturt University.

Leonardo Boff, 2015, *Come Holy Spirit: Inner Fire, Giver of Life and Comforter of the Poor*, New York: Orbis.

Leslie Boseto, 1995a, 'God as Community—God in Melanesian Theology', *Pacific Journal of Theology* 13:41–48.

———, 1995b, 'Do Not Separate Us from Our Land', *Pacific Journal of Theology* 13:69–72.

Tui Cadigan, 2010, '*Tangata Whenua*, People of the Land', in Dennis Gira *et al.* (eds), *Oceania and Indigenous Theologies*, pp. 60–5, London: SCM Press.

Pope Francis, 2015, *Encyclical Letter Laudato Si' of the Holy Father Francis on Care for our Common Home*, NSW: St Paul's Publications.

Gregory of Nazianzus, 1978, 'Orations 40.41', in Philip Schaff and Henry Wace (eds), *A Select Library of Nicene and Post-Nicene Fathers of the Church* vol 7, 2nd series (1893), Grand Rapids: Wm B. Eerdmans.

Winston Halapua, 2010, 'A Moana Rhythm of Well-Being', in Elaine Wainwright *et al.* (eds), *Spirit Possession, Theology, and Identity: A Pacific Exploration*, pp. 100–4, Hindmarsh, SA: ATF.

'Epeli Hauofa, 1993, 'Our Sea of Islands', in Eric Waddell *et al.* (eds), *A New Oceania: Rediscovering our Sea of Islands*, pp. 2–16, Suva: University of the South Pacific.

Pacific Church Leaders, 2017, *Sowing a New Seed of Ecumenism: Statement of Basis and Resolution*, Nadi: Pacific Conference of Churches.

Raimon Panikkar, 1993, *The Cosmotheandric Experience: Emerging Religious Consciousness*, New York: Orbis.

Tevita Mohenoa Puloka, 2007, 'In Tonga, a Straight Line is a Curve: Poetry as Metaphor is a Tongan Theology of Conservation', *Pacific Journal of Theology* 38:59–69.

Joerg Rieger, 2018, *Jesus vs. Caesar: For People Tired of Serving the Wrong God*, Nashville: Abingdon.

Jenny Te Paa Daniel, 2020, 'He Moemoea mo Oku Mokopuna', in James Bhagwan *et al.* (eds), *From the Deep: Pasifiki Voices for a New Story*, pp. 16–22, Suva: Pacific Theological College.

Amamalele Tofaeono, 2000, *Eco-Theology: AIGA – The Household of God, A Perspective from Living Myths and Traditions of Samoa*, Erlangen: Erlangen Verl. Fur Mission und Okumene.

Tui Atua Tupua Tamasese Efi, 2018, *Su'esu'e Manogi: In Search of Fragrance*, Wellington: Huia Press.

Sevati Tuwere, 2002, *Vanua: Towards a Fijian Theology of Place*, Suva: University of the South Pacific.

Upolu Lumā Vaai, 2019, *We are therefore we live: Pacific Eco-Relational Spirituality and Changing the Climate Change Story*, Policy Brief No.56, Tokyo: Toda Peace Institute.

———, 2020, 'Relational Theologizing: Why Pacific Islanders Think and Theologize Differently', *Pacific Journal of Theology* 58: 40–56.

——— and Aisake Casimira, 2017, 'Introduction', in Upolu Lumā Vaai and Aisake Casimira (eds), *Relational Hermeneutics: Decolonizing our Mindsets and the Pacific Itulagi*, pp. 1–14. Suva: University of the South Pacific and the Pacific Theological College.

Mark Wallace, 2019, 'The Stones will Cry Out: Christian Animist Crossings of the Species Divide', *Kosmos*, Summer 2019 (accessed 20.3.21: www.kosmosjournal.org/kj_article/the-stones-will-cry-out/).

19

out of darkness

Before the first dawn, darkness was deep and alive
before the first morn, darkness was the mother of life
before day one was born, darkness was the home of light

out of darkness, burst out light
into darkness, light reclines
with darkness, light journeys
in the skies, across lands and seas
and even in the cracklings of leaves

darkness is gentle and it incubates limbo'nations
it has space for the surplus populations
room for orphans of the digital dispensation
and patience for the disposable generations
its arms are wide, to embrace victims
vulnerable, sick and sickened
black and brown
even the grey and the white
and a host of shades all around

darkness speaks of God, abundantly
and in abundance, God embodies darkness
darkness allows visions to appear
and de-visions the faces of fear
darkness laughs at the fakes of rationality
and cries at the mentality of scar-city

darkness is daring – the bread of solidarity
and waters spiritual eyes – the current of diversity
it zooms across borders and over nations
but it is not limited to them with occupations
covid might be among revealers and teachers
but darkness follows the footsteps of tears

pierce the skies and let langimālie testify
that creation was made with the substance
of water, dirt and the sublime
seek darkness and find life in dirt and dust
walk humbly with the fragile and don't
be obsessed with the illusions of the divine

after the first dawn, darkness was whitewashed
after the first morn, darkness was lynched in life
after day one was born, darkness was the enemy of light
and then day two dawned, out of darkness

Jione Havea
Wurundjeri land & waters
Kulin nations
1 November 2020

20

Coronavirus Cacophony:
When the Dwarf Rebukes the Giant

JAMES W. PERKINSON

Zachary Bush (MD) recently asserted that the novel corona-virus leaping from winged cave dweller to human struggler is not so much killer as revelator – revealing humans' monstrous elaboration of toxic eco-destruction (Bush 2020). The essay sits before this subtle reframing to craft an echo-response to the virus as what popular writer Charles Eisenstein (2020) called a 'coronation' – a globe-wide initiation into a new demand of evolutionary 'sovereignty'. While both evocations stir cogitation and flicker with vision, neither takes its insight fully into the netherworld depths of our current dystopia. The writing will track these 'thought-comets', arcing over the fierce upheaval of now, as the effort of one white male settler at the curve of water called Detroit – wrestling with the provocative, but not-quite-sufficiently-critical inspirations of two other such white usurpers of place in the U.S. of A. – to push towards a more indigenously-responsible exploration of this new cross-over from beyond human self-preoccupation.

Certainly, the virus is a messenger – in its hypoxic invasion of human bodies, unmasking, at least in part, the air pollution whose urban vehicular or rural glyphosate-laced particulates become the combinatory laboratory that ups the ante on mortality (Bush 2020). Thus industrialized metropoles and industrialized agriculture stand revealed as lethal and damnable. But missing from Bush's breakdown of the disease profile is the physiognomy of its *social* apocalypse – unveiling the ongo-ing decimations that white supremacy and settler colonialism, neoliberalism and petro-capitalism continue to visit on populations indigenous and of colour. Eisenstein likewise embraces the microbial shapeshifter as teacher but fails to delineate the outrage of the asymmetry disclosed. How might we 'hear' the message – not losing or diminishing the wonder of wild nature's Gargantuan Potency,[1] packaged so minutely, in this late hour of the microbiome's signal role in ever 'looking for' adaptive redress to the grave unbalancing that larger 'players' like modern human social orders so wantonly foment on a now global ecological scale? But at the same time, not lose a single scintilla of prophetic indignation and denunciation of the oligarchic obscenity that the RNA-trader is one more time exposing *inside* human social relations themselves (what Paul tries to bring into focus as 'Principalities' and 'Powers' and 'Thrones' and 'Dominions'; Eph. 3:7–13; 6:10–17)? That is to say, how do we speak of sacrality in a manner not excluding the ironic 'majesty' of the nano-scale of Divine Numinosity while cross-cutting the amazement with a commensurately fierce exposé and challenge of the demonic?

Awe and Anguish

We begin with astonishment. What a strange time this SARS-CoV-2 has unleashed. A dwarf has taken down a monster, with an entire world watching. But what have we witnessed in this 'crown visitation' of nano-insurrection, coming over from the wild side? Our media casts the last six months of 2020 as neo-liberal catastrophe and declares war. One colleague of my wife names the virus 'messenger', speaking on behalf of an entire biosphere, declaiming, 'You cannot be well if we are not well. And we are not well' (Flyntz 2020). A Yoruban friend-scholar offers a sci-fi confabulation of fantastic peroration in which the little one – like all else – is finally unnamable, an Eshu-trickster shapeshifting through pangolin-creature to become 'human theriomorph', opening a crack in the world-mastery of capital (Akomolafe 2020). It is certainly patent that not even medicine in our day knows the exact frame and form. Is Covid a disease of the lung? A cardiovascular interruption of the cellular 'covenant' between air and blood? A provocateur of the virome, pointing away from itself, while irresistibly underscoring our penchant to pollute whether in terms of Hubei urbanity or Iowan love of Round-Up?[2] Every day brings new consternation … and desperate political scramble to make the fear yield votes and power. And it is perhaps this last word that most provokes.

Whatever else the viral voyager may be said to be, it has certainly become a 'power' of the time. And here we edge towards suggestive meditation, informed by a suite of Pauline terms, mobilized by the apostle in his day to discern and expose the inner workings of the first-century Roman *imperium* promulgated across the Mediterranean theatre of extraction and enslavement. 'Principalities and Powers' monikers can pinch back a bit of the veil masking large-scale spiritual predation that otherwise comports with brute material immediacy. In the flash of such a speculum, Covid can be partially tallied as a 'coerced' microbe translating white supremacist and settler colonial structure into neoliberal Violence and spiritual Delirium writ large across the globe. The viral microbe has been weaponized most devastatingly against bodies indigenous and of colour. But here we also need to back up. Paul's demonology, X-raying observable political hierarchy for its spiritual penumbra, laying bare a monstrous working not observable to the naked eye, itself demands accounting. What is a Principality? And whence?

Paul's language references city-state reality. But 'Thrones', 'Dominions', 'Principalities', 'Authorities', 'Dynasties' have not always been with us. They convene a tradition scarcely three millennia old when Paul takes up his pen. Before that, humans lived smaller-scale, with greater survival skill and flexible repertoires of response to seasonal cycle and watershed variability. The advent of the city-state in human history marks an upheaval (Scott 2017, pp. 1–35). Until the first appearance of such roughly 5,000 years ago in Mesopotamia (and later in China, Egypt and a few other hot spots around the globe), human cohabitation with non-human communities of plants, animals, waters, soils and weather was roughly egalitarian and committedly symbiotic and mutual (Gowdy 1998, pp. xiii, xv–xxxi; Lee 1998, pp. ix–xii). Whether in hunter-gatherer bands, horticultural villages or pastoral nomad clans, kin relations defined interaction and goods were largely shared. Neither cultural practice nor narrative code tried to underscore human 'being' as in any sense special or supreme (Kimmerer 2013, pp. 176–7). They rather mandated a broader spectrum of identification, typically bounded by the local bioregion, inclusive of many non-human actors who altogether made life possible (Prechtel 2012, pp. 5, 8, 49–51; LaDuke 2005, p. 228). Biodiversity cut *through* humanity; personhood was predicated of much of the varied life forms; spirit was hailed as ubiquitous. To be 'human' was to be, at the same time, *more-than-human* (to be, that is, camel-people, oak-people, whale-people) (Corbett, 1991, pp. 4, 8, 85, 88; Haraway, 2015)! And given the frequent mobility, lack of settlement density and seasonally respectful food-web recursivity, disease outbreak was virtually unknown. Pandemics did not yet exist in history. Animal habitat was not yet under pressure and zoonotic cross-over not yet even a glint in the viral 'eye'.

The city-state changed that. While the shift to agricultural regimes of coercively constrained labour, forced to generate surplus product for consumption and accumulation, took place for reasons perhaps related to climate change, but not yet fully understood – the shift was epochal (or better, cataclysmic). All of the early city-states were effectively slave machines, regularly aggressing on nearby population centres (villages, towns or competing city-states) or peripheral areas (mountains, marshes, grasslands) to replenish labour forces that continually fled (Scott 2017, pp. 153–4, 193). Work drudgery, imposed debt, and recurrent disease epidemics were the labouring lot of such urbanized polities (Scott 2017, pp. 47, 117, 121, 129). And in the crammed-together impoverishment of domesticated animals made to serve domesticated humans in cultivating domesticated plants, zoonotic crossover began its irrepressible career in our species. And obviously, I am sketching in broad brush here. I will go even broader.

With all of our modern scientific proclivity and hi-tech dexterity, we still do not know what a virus 'is'. The same can be said for virtually everything. 'Identity' is largely a function of drawing artificial boundaries with words and then falling for our own conceits (Haraway 2015; Akomalafe 2020). Nothing exists by itself. All bodies are continuously being added to and subtracted from by other bodies (what we call 'feeding' and 'defecating' among animals, 'fungal sheathing' and 'photosynthesis' for plants, 'oxidation' and 'erosion' undergone by rocks and minerals, etc.). With viruses, however, we cannot decide whether to christen them as 'living' or 'inanimate'. What *does* appear extant, however, is the advent of these strange little wiggly-squigs as the original exchange artists, 'multicultural proliferators' extraordinaire – from way back before Genesis-time, with RNA-quixotism, hijacking DNA from here and trading it there, birthing biodiversity straight from the loins of homogeneity. They are our ancestors among other mantles they carry. And they ride in (and 'as') our bodies today to the tune of some 380 trillion per person (Pride and Ghose 2018).

Zach Bush, mentioned at the start of this chapter, insists that these questing viral invaders be themselves apprehended in ensemble with the entire microbiome as evolutionary voyageurs, in momentous up-wellings of apocalypse, arcing out of overstressed predecessors, to *adapt* the whole project of 'life' to whatever brave new world is incubating. Viruses then are that through which all else living was made! Shaman-dwarfs, shapeshifting the continuum of life into ever proliferating diversities, questing for the right fit! At one level, the most massive communication exchanges over three billion years of animate confabulation on this blue marble have been the continuous 'tweets' between microbes and genomes – between viruses and bacteria and DNA, that is – in comparison with which, human communication over three million years is but a blip on the screen.

Thus, in the beginning, we might say (only half in jest), was the virome – the original *Demiurgoi* that Plato imagined to be but 'One', Hebrew scribes hypostasized as Lady Wisdom (*Hokmah*), and John the Gospeller pontificated as the *Logos*-Word! Here is 'code' crossing over the thresholds of incarnation, 'speech' leaping from the Other World to this one, 'tenting among us', in the flesh! Though invasive, as pirate-ambassadors of difference and provocateurs of innovation, their outbreaks served to facilitate life.

Symbiosis or Surplus?

But in early city-states, these information-traders also began to be weaponized as Powers. No longer merely adventitious pioneers across the boundaries of eukaryote cell-walls, viruses found increased virulence as part of the gathered force of coerced labour. Political scientist James C. Scott summarizes the import: the first city-states were incubation chambers for infection, in-walled work-regimes answering to the earliest impulses of technologies of control (Scott 2017, pp. 18–19, 21, 31, 93ff.). Far from some inevitable outcome of an imagined progressive thrust of history, the turn to concentrated city-state organizations 'between the rivers' (in 'Mesopotamia') some-where around 3000 BCE (and later elsewhere) required constant coercion (Scott 2017, pp. 25, 27, 29). No hunter-gatherer or multi-skilled horticulturalist/pastoralist would have voluntar-ily abandoned a lifestyle of relative leisure and autonomy for hard labour and tax-gouging as a virtual slave of such a state (Scott 2017, pp. 8, 18, 22, 88–9). They had to be coerced into the regimen in the first place and controlled by taxation, food rationing, military conscription, corvée imposition, and walls and overseers, as well as ideology, once there (Scott 2017, pp. 29, 137, 150ff.). But control admits of nuance and, in and of itself, is not the bogeyman.

Collaborative control, control committed to a shared flourishing, control balanced with unpredictable innovation in service of a wild efflorescence of biodiverse itineraries of being, would seem to be part of the regimen of life. Geologist Marcia Bjornerud describes the mysteriousness of a phenomenon observed in local ecosystems across the planet, wherein some overarching force-field of selection has maintained biomes in rough balance for some half billion years (Bjornerud 2005, pp. 92–5). Species within niches in such an eco-community come and go with great frequency, but the overall structure of the niches in relationship to each other – smaller bodied consorts outnumbering larger bodied occupants in trophic proportionalities commensurate with overall viability – remain relatively stable. Population numbers inversely related to body size were regulated by eating relations. This would seem to be a given of 'life', a biodiverse mandate maintained over time by the gestalt of biomic mutuality. And within this shared regime of trophic scaling, the primary producers, on top of which all the ascending hierarchs of herbivores, insectivores and carnivores find their possibilities for thriving in descending numbers proportionate to increasing body size, are single-celled autotrophs, especially in the form of bacteria, drawing energy from the sun (via photosynthesis) and rock and spring by way of microchemistry (Bjornerud 2005, pp. 93–4). Bacteria thus emerge as the prime movers on the planet. And viruses have effloresced as the prime distributors, adapting bacterial production to changing biomic conditions and such conditions to changing bacteria.

But Bjornerud also notes an anomaly (p. 95). Humans by way of agricultural technologies over the last five millennia would seem to be on quest to exempt themselves from this half-a-billion-year-old synergy of trophic scaling laws – far exceeding appropriate numbers assigned by body weight, gobbling up the energy production of their co-dwellers in gargantuan appetite, obliterating species after species.[3] The conceit has been that it is all there 'for us' – as so much of human supremacist thinking would have it. Whoever can take it deserves it. Five thousand years, however, is but a blip in the outworking of evolution. The destiny of such a presumed exemption appears, in our day, indeed presumptuous. And likely doomed. And just here, the settled-agriculturalist city-state as the taken-for-granted engine of civilization begs inquiry and challenge. It represents a departure from the laws and regimens and control pressures of 500 million years of emergent co-habitation. Arguably, its problem is indeed one of scale. It represents a rupturing of scaling laws in the direction of inflation of the role of certain species at the expense of others.

To wit, for instance, wheat (or any other monocrop favourite of urban settlements such as barley, rice, millet, corn, etc.).
Monocrop agricultural development in the ancient near east
abandoned early flood retreat reliance on wild example (or controlled burning in other places on the globe) to begin wholesale
rearrangement of the environment by clearcutting forests, irri-
gating floodplains, damming tides, draining bogs (Scott 2017,
pp. 20, 55, 66, 70, 121–3). A crop of choice such as einkorn
wheat would be planted far in excess of its naturally occurring
presence, while all competitor plants were regularly 'de-selected'
(pulled up, weeded out, etc.) (Scott 2017, pp. 73–5). Thus,
wheat was fated and 'inflated' to become an 'enslaved' domes-
ticate, bred to remain on the rachis to facilitate harvesting,
pushed into an outsized productive fruition, serving an out-
sized human population gathered in and around a city centre.
Wheat was remade a 'Power', in the biblical sense.

We are hardly used to thinking Paul's language innovation
this way. The 'Principalities and Powers' nomenclature Paul
deployed served to delineate a deciphered inner capacity of
Roman imperial infrastructure that otherwise remained invisible in its effects (Wink 1986, pp. 4–5). Paul drew on Jewish
adaptation of ancient (Mesopotamian, Egyptian, Canaanite,
Zoroastrian, etc.) intuition of behind-the-scenes operators
bringing heavenly potency into play in earthly destiny and
events. 'Angels' became the watchword for such in later Jewish
and early Christian tradition, denoting spiritual messengers
regularly crossing boundaries between worlds to facilitate
communication-exchange and power-access. But when we
personify the potency, we can easily infantilize the conception.
In developing Christian tradition, the angelic realm is arranged
in a hierarchy of function, often correlated with astral designation (Angels, Archangels, Principalities or Rulers, Powers or
Authorities, Strongholds or Virtues, Dominions or Lordships,
Thrones, Cherubim, Seraphim). By the high Middle Ages,
philosophy Islamic, Jewish and Christian will have predicated
angelic 'souls' (or 'motions') of each of the celestial spheres
they observed in the heavens and routinized the ensemble in a
melodic bureaucracy – with angelic choirs lauding the divine

throne, located higher up and further away. But the 'fall' was then given cosmological purchase as having encompassed not just humans, but part of this angelic realm. Lucifer supposedly careened 'out of position' down into terrestrial entrapment and vocation for having sought to elevate himself above his station (as cherubic in one arrangement) (Luke 10.18; Isa. 14.12–17). Other angels were imagined to have followed suit and the result was a new 'settlement' of angelic power in human institution, an incarnation of other-worldly force in this-worldly embodi- ment, named and declaimed in terminology largely political, all inflating beyond their legitimate vocation.

But it is crucial to noodle this schema in history. Paul is clearly pinning older conceptions on contemporary realities to render his realm at least 'darkly' decipherable (the mirror-words with which he is 'scrying' only catch sight of these behind-the-scenes Powers in ghostly outline; 1 Cor. 13.12). Principalities, Rulers, Powers, Authorities, Strongholds, Dominions, Lordships and Thrones all have their correlative material references in Roman political bureaucracy and pageant. But way back in the Hebrew tradition of Paul's own scribal formation, the language used was Nephilim, updated and amplified by the third- to first- century BCE writers of Enochian literature as 'Watchers'. And here begins to coalesce a veritable 'host' of reinforcing recog- nitions. The Enochian elaboration of the Genesis 6 invocation of an ancient memory of 'giant' Watcher Angels, 'falling' due to their inordinate love for human women and descending to earth to propagate a new race of *bene Elohim* ('sons of God' or 'angel-humans') is dim apprehension indeed. Much must be left up to conjecture but throbbing quite close to the sur- face is another association, underscored with particular savvy by French lawyer-theologian Jacques Ellul, to the effect that the Hebrew word for 'Watching Angel', *'iyr*, is also the word for 'urban settlement' or 'city' (Ellul 1970, p. 9). That the word also conveys resonances of 'terror', 'trembling', 'fear' and 'anger' specifies the kind of 'watching' implied. The city incarnated 'Dread Surveillance'.

Scott underscores the import even in the title of one of his works (*Seeing Like a State*, 1999). The city was above all a new discipline of 'scopic' control, looking down from high towers on its labourers and surroundings. The emblem of early monarchy was the plumb line – the technology of surveying land that marked its boundaries in visible media and incorporated all within its imposed jurisdiction under the regime of forced labour and imposed debt known as taxation. Census-taking emerged as the necessary correlate of tax-gathering, making otherwise inchoate populations transparent to the ruling class 'will-to-extract'. And wheat, barley, oats, millet (and elsewhere rice and corn) became the primary food crops of state organization because, unlike tuber staples (cassava, potatoes, yams) they were 'legible': they grew above ground, matured at roughly the same time, could readily be assessed and seized, and were relatively easy to transport and store and thus withhold to secure compliance or send out with troops on military campaigns (Scott 2017, pp. 21–5). City-states indeed materialized 'Watching Angels'. But the Powers so referenced were neither singular nor angelic. They amassed a collective potency that was outsized and 'out of place'.

Wild Potency or Urban Profligacy?

We back up to sketch from a different angle. The line of thought followed above takes off from recognition that Covid emerged as a 'pandemic' not because of its innate character as a virus, but because of the artificially constructed terrain into which it 'leapt' from bat caves or pangolin wanderings. The socio-economic and politico-cultural infrastructure of globalized urbanization that organizes and adjudicates most human interaction in most places on the planet today, has mobilized the virus as one more differential Potency among those already arrayed in a warped and unsustainable constellation of hierarchies. Unlike the trophic hierarchies of ever-changing biomes over half-a-billion years, human reorganization of energy systems and caloric exchange in service of elite accumulation over the last five millennia does not answer to 'natural' scaling laws. Wild nature, biomass, metals, minerals, trees, waters, weeds, pigs, sand, rock, dead mollusk remains from ancient seas – virtually everything extant on the planetary surface and increasingly 'creatures' long buried even as deep as one mile or more, are all being ripped from context and redeployed in service of an aggrandizement that is patently unsustainable. Water pipes, highways, fibre-optic cables, sky-scrapers, plastic bags, tanks, trucks, bombers, Pentagons, oil rigs, industrial farms etc. are all composed of matter violently pulled from an incalculably diverse patchwork of gift-economy ecosystems whose track record of survival long-term renders urban hubris over supposed accomplishment rip-roaringly laughable if it were not so utterly tragic and obscene. And all of the assembled plunder now articulates an emergent gestalt, a supervening set of competing Agencies, not entirely calcula-ble by simply adding up the parts.

When we turn to Paul to grasp an outline of the complexity we face, unwittingly we are also edging towards a more indigen-ous assessment. In the savvy of many archaic peoples, all things are at some level spiritual and living. Every ant or twig, rain-drop or rock outcrop carries something of the whole in which it is set. It is spiritually potent in a manner impossible to dissect. The spirit cannot be pried loose from the body. Each material element of the ecosystem is also aura-ed with spirit. Paul indeed pushes a bit in this direction in rooting his hierarchy of Powers in what he calls 'elemental spirits'. Behind-the-scenes spiritual forces not only show up at the level of obvious political coercions (Authorities, Dominions, Thrones, Principalities etc. in the form of armies, courts, emperors, magistrates, regalia), not only in the seemingly blatant manner of individual 'break down' (Demon possession exhibited as untoward behaviour), but in the dynamism of elementary particles of the world. And here the possible candidates are as manifold as the created order itself. Exegete Walter Wink maps the range from the micro-scopic to the gargantuan (Wink 1986, pp. 130–4, 141–4).

What Roman citizen Paul apparently sees when he looks at an imperial city of his day is a writhing mass of enslaved Powers, gathered from all over the Mediterranean Basin – wood, iron, copper, sandstone, granite, grain, horn, gold, amber and human

– now coalesced into a rampaging imperial juggernaut. And all of it throbs with a now shredded 'spiritual' energy. No longer part of well-tested and slowly evolving force-fields of scale-relations, each 'thing' in the ensemble has been effectively hacked out of its organic 'home' and 'hot-pressed' into an alien 'slave' machinery (ecologists today increasingly identify civilization as a doomsday heat-engine). Which is to say 'things' are now beset with an untenable simultaneity: they are both less and more than they were in wild provenience. No longer part of a magnificence regulated by scale into an astonishing epochal motility and consonance, they are at one level trivialized. At the same time, coerced into subservience of a 'Frankenstein' or 'Godzilla' enterprise run by insanely self-aggrandizing human elites, they are now also forced to goose-step in Algorithmic Monstrosity. Whether as Weapon or Tool, dominating Technique or insatiable Extraction, the resulting Titanism functions like nothing so much as a Hungry Ghost appetite in Full Metal Jacket armature, 'eating' a planet. Star Trek imagined the outcome as that of a planetary rotundity re-engineered into a Cy-Borg cube, hard-wiring into its mission of galactic conquest every species it overtook! Language here obviously strains at its inadequacy. But in Pauline coinage, we have a renegade multiplicity of fallen 'angelic' Powers, ranging from minute to monstrous, conflating and competing with one another in a win/lose game whose ultimate triumph is precisely ultimate loss (should they win in an ultimate sense, humanity itself would likely disappear as the living substrate of such a renegade potency).

Flourishing or Fracturing?

And this pushes towards our final demarcation. The Powers are plural and fracturing, driven by a one-up logic, pretending loss is expendable. Part of the impossibility of corporatized externalities today is the globalized emergence of a new disposition of 'homeless matter' as 'garbage'. Plastic alone registers the crisis: it is on course by 2050 to exceed fish as the most ubiquitous phenomenon in our oceans (Davis 2020). E-waste as well is mushrooming like a metastasizing disease (not to mention material like fuel rods and radioactive debris) (Singer 2020). But it is not clear if 'trash' actually existed before urbanized settlement. 'Compost' was the predecessor 'collectivity'. In a rhythm natural and organic, this particular eco-extravaganza magically transformed defecation into food, refuse into renewal. And no surprise then that a number of indigenous cultures hallowed the phenome as divine, the 'goddess of failure' making possible all other conditions of life (in the case of the Mayans, actually erecting shrines over the compost piles in a sacred entombment that colonial anthropology would later mistake for tomb-raiding detritus) (Prechtel 2012, pp. 10–11, 33, 47–8 and various in-person talks).

Gospel hints about Jesus Movement spirituality partake of a somewhat similar orientation, though hidden in plain view and perhaps strange in association. Confronted with individual lives broken into aberrant behaviour, the Nazareth prophet might well 'exorcize spirits', seemingly banishing the invading 'infection' elsewhere, like discarding 'junk'. But when faced with the institutional version of such spirit-possession – a Principality or Power in the Pauline language – his tactic is not exorcism but downsizing – what we might recast as spiritual composting. It is worth spending an exegetical moment with the paragon passages. After inaugurating his ministry in Capernaum in Mark's account with a 'send up' exorcism – clearing a plurality of spirits from a beleaguered body and indeed, from the synagogue space itself such that its silencing ethos is punctured and even peasant small farmers suddenly feel entitled to speak and question – Jesus goes briefly feral, up a mountain, to solidify his inner circle and his movement aim (Mark 1.21–28; 3.13–19a). Upon descent back into public accessibility in Capernaum, he is immediately accosted by Jerusalem surveillance and disinformation, sent out from the Temple-State centre to begin the campaign that will secure his defamation and 'disappearance'. The charge is scabrous. His own Spirit-Power is supposedly nefarious. He casts out demons by Beelzebul, the Prince of Demons. He is a cog in a much larger operation, beholden to a Force, Fallen and Sinister in the extreme.

His response is startling. He does not deny the accusation, but instead 'riddles' the charge with inquiry. He calls the village crowd to lean close and think, hard. It is a teachable moment – precisely about the hierarchy the scribal spies have introduced. Just how does Spiritual Power in its harder-to-track institutional insinuations, manoeuvre? And what is its hallmark? Going straight for the jugular, he grills: 'How can Satan cast out Satan?' But then sidesteps into parabolic innuendo: 'If a kingdom is divided … if a house is divided?' Suddenly we are no longer abstract and theoretical but beginning to dance with the real devil. 'Kingdom' immediately brings the actual political power over that fishing hamlet of Capernaum into the mix: they are in Herod's murderously oppressive domain. And 'house' invokes David's line centred in the Temple-Shrine in Jerusalem, whence the interrogators had come and where power's economic and religious instrumentalities such as records of indebtedness and tithe-arrearage are kept and leveraged. Now we are getting into the real meat. Large-scale Spiritual Power works through banal policy and exclusionary pageantry. Its means is divisive; its goal extractive; its outcome destructive. The question on the table is indeed one of Principalities. But the parabolic hint is that it is the surveilling scribes voicing the charge who are the real agents of Netherworld malevolence. Will the real Fallen Angel please show itself! Jesus will conclude the fraught polemic with bald assertion: 'You cannot plunder a Strong Man's goods unless you first bind the Strong Man.' And herewith the subtle indigenous valuing! The tack will be 'binding', not 'exorcism'. But Mark's account is cryptic; we need to avail of elaborations by Luke and John to catch the full import.

In Luke's version of the event, Jesus is tagged with the Beelzebul charge and explicitly 'tested' (Luke 11.14–23). After running through his 'kingdom' and 'house' counterattacks, he elaborates. The 'fully armed' Strong Man can only be overcome by One 'Stronger', who is able to constrain the former, taking away his armour, and dividing his spoil. But this is division with a difference – not the creation of winners and losers, wealthy dominators and enslaved labourers in an extractive system, incarnating hyper-inflating 'Powers', but a disaggregation and recirculation of coerced elements back into something more like gift-economy mutuality, reciprocal exchange. The Greek is explicit: the 'panoply' of defences is *dismantled*, not destroyed; the goods *redistributed*, not hoarded or banished. Otherworld healing here does not truck in 'winners and losers', or in 'spiritual garbage heaps'. Rather, we can extrapolate. The battle is finally one of scale – of returning the pirated infrastructure and assembled armature of empire back to their more natural and sustainable 'vocational niches' as one or another physical incarnation of 'spirited' matter.[4]

The Revelator John continues the intuition. In his 'Apocalypse of Babylon' riff (chapter 18) concluding his 22-chapter-long 'end of the Roman Empire' vision, the details of the 'fall' of the archetypal city are instructive. While certainly worthy of critique in its gender representations, the picture presented gives a litany of trade materials 'composing' the monstrosity in its wanton assemblage of wealth and hubris (Rev. 18). The list of linens, gems, metals, woods, ivory, minerals, spices, oils, flour, cattle, horses, chariots, wheat – and yes, human 'slaves' – is the concrete materiality of the city finally summed up in the charge: 'And in [it] was found the blood of prophets and saints, and of all who have been slain on earth' (Rev. 18.24). X-ray vision of any city would disclose the same genealogy – venerated bodies of matter that in reality are soaked in blood! Urban architecture in visionary necropsy contains most of the violence of history. And then juxtaposed to such a horrific unveiling is anticipation of the 'new Jerusalem' descending from the galactic immensity like a meteoritic 'star', full of healing leaves, crystalline streets, living waters, bejewelled, 'angel-endowed' gates (Rev. 21.12), thunder-throned (Rev. 19.6), sea-washed, and a height equal to its length and width – which is to say a physicality of urbanity unlike anything 'urban' ever constructed by humans. The term 'city' for this emergent novelty of physical enshrinement and sacrality is at best approximate, but in truth, beyond anything the word can conjure. And the telltale conceit here – its 'measure' is that 'of an angel's' (Rev. 21.19)! So yes, the city as the 'house' of angelic reality, whether in fallen form as inflated and blood-soaked Powers and Principalities, or in honourable form as stellar, mineral, crystal, tree-growing, and flowing with 'alive' water which is also declaimed, as it once had been both by Isaiah (55) and Jesus (John 7.37–40) as 'living Spirit' (Rev. 22.17). And we circle back to where we began.

Not least among Babylon's natural scourges (fire, famine, pestilence) 'repaying the city double for [its] deeds', is plague (Rev. 18.4). It is striking that of the haunting *spirits* enumerated that belie its mesmerizing facade of riches, one is a lamented infestation of 'foul and hateful' (from the point of view of the city itself) birds (Rev. 18.2). Of course, we could throw in 'vermin' of all kinds, the 'sparrows, mice, weevils, ticks, bedbugs' that Scott calls euphemistically, but accurately, 'commensals' – all of the crawling and winged opportunists for whom city life is a gathered fest of 'good feeding'. These are part of what is gathered 'out of scale' into the armature of the Watching Powers presiding over human institutional pretension, the wild ones weaponized in violation of their natural disposition, against mutual flourishing. There is also here an inverse effect: the smaller-bodied decides the fate of the larger, the 'dwarf-beings' rebuke the big ones grown monstrous in their refusal of limitation and 'place'.

Peril or Prophet?

And thus is Covid. What is perhaps untoward in my telling here, not a little obscene and scandalous for the developmentalist mind as for the evangelical absolutist, is the idea that the virus may well be angelic. Certainly, the poets have it right that it be assessed, among other characterizations, as 'messenger'. This is but the actual import of the word *angelion* in Greek – a go-between informant, crossing world-thresholds to convey news and warnings and heretofore uncalculated pressures on events.

In Pauline assessment, run through Wink's exegesis, not just more obvious anthropomorphic actors captured in stained glass figures, but much more inchoate influences such as scale laws, force-fields (such as electromagnetic or nuclear or gravitational

'Powers'), fractal proportionalities, propensities of bodies in motion to synchronize rhythms, symbioses not clairvoyant to human eyes, viral genetic exchanges, etc. might be comprehended as the workings of behind-the-scenes Forces otherwise 'collectivized' as 'Elemental' Angels (Wink, pp. 130–2, 143–4). In his Apocalypse, John did not address humans; he wrote rather to the 'group ambiance' known as the 'angel' of the church of Laodicea or Ephesus or Philadelphia (Rev. 2.1; 3.7, 14). An ecclesial gathering – as any other integral collectivity on the planet, human or other – is spiritually more than the sum of its constituents. But so also are non-integral entities, gargantuan assemblies, out-sized totalities. Principalities and Powers in Tyrian palaces as indeed Dominions in Roman courts or Cherubim in Jewish Temples or Strongholds in Mediterranean shipping enterprises are 'creatures' that Paul both 'contends against' and seeks to make 'wisdom known to' (as if, in fact, they might be 'recaptured' and returned to appropriate scale; Eph. 6.12; 3.10; 4.8 riffing on Ps. 68.18). Which of these more-than-meets-the-eye-coherences the virus known as 'corona' ends up most serving, rests, not with the virus but with us.

There is no little cautionary irony in the designation 'corona', given Christian medieval artistic depictions of angelic hierarchy. The Power represented as 'crown-wearing' in the eyrie assemble is precisely that of the Principality. Thus far in our disparaging and deaf approach to the messenger, Covid has indeed been made to function as a devastating Beelzebulian 'Prince'. But there is yet a tiny possibility we could still choose to embrace the nano-being as prophet rather than peril. If so, the council is irreducibly radical. It is the entirety of our urban-industrial petro-aggression on the biosphere that now stands revealed as demonic. Are we even capable of 'owning' what is being unveiled, much less mobilizing to change? It does not look good.

Notes

1 Throughout the piece, I off-and-on capitalize references to natural and wild phenomenon to emphasize a more indigenous perspective in

which such 'beings' are narrated as 'living', 'agentive' and 'spiritual'. I recognize that these terms ('wild' and 'natural') are not innocent of settler colonial history and can eclipse indigenous presence from the landscape and the discourse. But I do want to talk (somehow) about the Planet and Her creatures, at times, in their existence, priorities, rights, influence, trauma, etc. separate from our own concerns as perhaps the youngest species coming into existence here, and do so under the influence of a major indigenous teaching presence in my own life, Martín Prechtel.

2 Bush wonders to what degree the virus camps out on poor air quality as an exacerbation of what is already profoundly toxic.

3 Indeed, the extinction rates we have precipitated across the planet (200 species per day according to some measures) are being matched by an equally calamitous interior 'massacre' and extinction-event. As explored by the American Gut Project, the microbiome of modern Westernized human bodies is only 10 per cent as diverse as that of the hunter-gatherer Khoi San of South Africa, whose robust viral endowment derived from the natural wild environments they dwell in, gives them much more resistance than we currently carry, but immediately begins to reduce, homogenize and lose resilience as soon as the foragers are forced into settled living and crop-growing.

4 We could go further: here may be the basic intuition of the word 'religion', pointing towards the necessity of a *re-ligare*, re-ligaturing, re-binding what had been loosed from mutual flourishing to aggrandize destructively. In this sense, indigenous cultures do not practise 'religion' per se; having never 'untied' the big Powers of nature, but rather honoured them 'in place', they had no need to focus concerted attention on returning them to the bonds of gift-economy reciprocity. Here also is a hint for liberation theologies writ large: 'freedom' is never absolute, but a release *from* oppressive relations of domination back *into* mutually 'binding' ones, of mutual care and limitation.

References

Bauyo Akomolafe, 2020, 'I, Coronavirus. Mother. Monster. Activist' (accessed 22.3.21: https://bayoakomolafe.net/project/i-coronavirus-mother-monster-activist/).

Marcia Bjornerud, 2005, *Reading the Rocks: The Autobiography of the Earth*, pp. 92–5, Cambridge, MA: Westview Press.

Zachary Bush, 2020, 'The Highwire: Evolutionary Virus Discussion', *Highwire Interview with Del Bigtree*, 18 May (accessed 22.3.21: https://zachbushmd.com/video/the-highwire/).

Jim Corbett, 2005, *A Sanctuary for All Life: the Cowbalah of Jim Corbett*, Englewood, CO: Howling Dog Press.

———, 1991, *Goatwalking*, New York: Viking Press.

John Davis, 2020, 'Global Plastic', *OpEdNews*, 2 July (accessed 22.3.21: www.opednews.com/articles/ Global-Plastic-by-John-Davis-Biomass_ Blowback_Chemical_Consumer-200207-768.html).

Charles Eisenstein, 2020, 'The Coronation' (accessed 22.3.21: https:// charleseisenstein.org/essays/the-coronation/).

Jacques Ellul, 1970, *The Meaning of the City*, trans. D. Pardee, Grand Rapids: Eerdmans.

Kristin Flyntz, 2020, 'An Imagined Letter from Covid-19 to Humans', cited in G. A. Bradshaw, 'A Message From the Virus, or Our Ancestors?: Nature is giving us another chance', *Psychology Today*, Mar 24 (accessed 16.7.20: www.psychologytoday.com/us/blog/bear-in-mind/202003/message-the-virus-or-our-ancestors).

John Gowdy, 1998, *Limited Wants, Unlimited Means: A Reader on Hunter-Gatherer Economics and the Environment*, Washington, DC: Island Press.

Donna Haraway, 2015, 'Anthropocene, Capitalocene, Plantationocene, Chthulucene: Making kin', *Environmental Humanities* 6.1: 159–65.

Robin Wall Kimmerer, 2013, *Braiding Sweetgrass*, Minneapolis: Milkweed Editions.

Winona LaDuke, 2005, *Recovering the Sacred: The Power of Naming and Claiming*, Cambridge, MA: South End Press.

Richard Lee, 1998, *Limited Wants, Unlimited Means: A Reader on Hunter-Gatherer Economics and the Environment*, Washington, DC: Island Press.

Martín Prechtel, 2012, *The Unlikely Peace at Cuchumaquic: The Parallel Lives of People as Plants: Keeping Seeds Alive*, Berkeley, CA: North Atlantic Books.

David Pride and Chandrabali Ghose, 2018, 'Meet the trillions of viruses that make up your virome', *The Conversation*, 11 October (accessed 22.3.21: https://theconversation.com/meet-the-trillions-of-viruses-that-make-up-your-virome-104105).

James C. Scott, 2017, *Against the Grain: A Deep History of the Earliest States*, New Haven, CN: Yale University Press.

———, 1999, *Seeing Like a State: How Certain Schemes to Improve the Human Condition Have Failed*, New Haven, CN: Yale University Press.

Katie Singer, 2020, 'Behind the Screens: The True Costs of Internet Access', *SciTech*, 27 April (accessed 22.3.21: www.opednews.com/articles/Behind-the-Screens-The-Tr-by-Katie-Singer-Activism-Envi ronmental_Climate-Change_Computers_Energy-200427-574.html).

Walter Wink, 1986, *Unmasking the Powers: The Invisible Forces that Determine Human Existence*, Philadelphia: Fortress Press.

21

Not Returning to the Old Normal

ANTHONY G. REDDIE

Systemic racism existed before George Floyd's brutal murder in 2020. For centuries, racism has been a part of the body politic of the UK and indeed of the world. In this chapter, I argue that if we want to be ruthlessly committed to eradicating racism and effecting a modicum of racial justice especially within our churches, then we need to address some of the historic, perennial problems that lie within our history.

The Roots of the Old Normal

The roots of the old normal lie in understanding the historic role white Christianity played in the Transatlantic slave trade. I am arguing that there existed (and continues to this day) an underlying framework that enabled many Christian churches to construct an ideology, based upon an incipient, racist theology that assisted them in supporting black chattel slavery, which was unhindered by any faith in God.[1]

To understand the churches' role in slavery we need to go back to the early Church Fathers and ideas derived from Greek Antiquity. It is in this much earlier period in the first four centuries of the common era that ideas of black people as the negative 'other' first begin to surface in Christian thinking. The later period of European Expansion around the time of the Crusades and the violent conflict with African (black) Moors (Muslims) leads to the intensifying of ideas around Christianity = Europe (Christendom) = White *versus* Non-Christians = Africa (Barbarians) = Black. Black people became the other (Hood 1994).

The aforementioned is exacerbated by the fact that White Christianity *is a violent religion*. It is based upon a form of 'closed monotheism' – i.e., the 'Christian God' is a jealous and competitive God that will not tolerate rivals and the 'other' that worships such God(s) (Douglas 2005) – which in turn is conflated with white exceptionalism, privilege and power. So, the conflation of white Christianity and the hermeneutics of power leads to forms of aggressive social-political praxis that is often predicated on violence. This can be seen in a number of Hebrew Bible texts, in which a 'competitive' God instructs the people of Israel to commit genocide on others who inhabit the 'Promised Land'[2] (see Exodus 23.20–33 and the list of peoples overthrown in Joshua 12).[3]

In invoking the term 'violent religion' for white Christianity I am speaking towards the wider Judaeo-Christian tradition which, when allied to notions of white supremacy, becomes the hermeneutical lens for rereading the texts in Exodus and Joshua, on which 'Christian genocide' is enacted. This view is explicated in the work of Robert Warrior (1997), reflected in the manifest destiny of white settler communities in the US, whose use of the 'closed monotheism' of Christianity enabled the justification for the usurping of Native American land. Warrior's claim to the violent impulses of white Christianity that has helped to fuel the white supremacy that underpinned European imperialism is amplified in the work of a number of international scholars and activists from the global South. These scholars and activists have demonstrated the means by which white Christianity has been able to colonize the Judaeo-Christian tradition in order to dominate and subjugate others, often people of darker skin across the world (see Hopkins and Lewis 2014).

When you combine the questionable attitude to blackness with the sense of competition with people who are not like you (i.e., black), you have a potent cocktail for an underlying theology of 'them' (black people or 'other') and 'us'. You know who the 'them' are, because they do not look like you. They are not 'of God' (not 'his people') and therefore, 'all bets are off' in terms of how you treat them.

When European traders, particularly in the Elizabethan age, begin to engage with Africans on a prolonged basis, it did not take much imagination to see that the underlying notions of 'otherness' made black Africans ripe for exploitation (see Gerzina 1995). The tensions between religion, faith, ethnicity and nationality are then exploited by means of 'specious' biblical interpretation. The main text that resolved the issue of justifying the enslavement of Africans within a Christian framework was Genesis 9.18–25 – the curse of Ham. Noah punishes his son Ham by cursing his own grandson Canaan (the son of Ham), condemning him and all his descendants to slavery (see Johnson 2004). Since there was a widely per-petuated belief that Africans/dark-skinned peoples were the descendants of Ham, this so-called curse of Ham was used as biblical evidence that the enslavement of African people was actually willed and sanctioned by God. There was also a similar but less well-known argument based on the biblical story of Cain and Abel (Gen. 4.8–16) where the 'mark of Cain' (punishment for the murder of his brother) is interpreted as representing black skin. Again, people of African origin are somehow identified as cursed by God for some past wrong. Here, any notions of blame are removed from the slave owners since it can be said that the condition in which the Africans find themselves as slaves is due to the sins that their ancestors committed in the past, for which God is punishing them. Their black skin is seen as proof of their sinful condition (see Hood 1994). Proponents of the Atlantic slave trade constructed such wild and fantastical forms interpretation of the Bible (in sup-port of slavery) because of the presence of pre-existing views of Africans as 'other' and as being 'cursed by God' (see Douglas 2005).

The aforementioned is ameliorated after the Haitian revolu-tion at the end of the eighteenth century. The charge to 'Christianize' enslaved Africans is undertaken on a number of biblical and theological terms. There is a dichotomy between the body and the soul (an outworking of Pauline theology) and salvation is achieved solely by faith in Jesus Christ. In Pauline theology, salvation is not dependent on praxis but on faith in the saving work of Jesus. This means that if you are a Christian, you can have faith in Christ and still own slaves, as God is only interested in your soul, which is preserved through faith in Jesus. Your actions on earth are another matter (Douglas 2005, pp.150–98). For the enslaved Africans, faith in this same Jesus guaranteed salvation in heaven but not material freedom on earth for the same reason as that given for the justification of slave masters. In the theological construction of slave holding economies, Africans could be saved. Given that this underlying framework of European superiority still held sway, however, even when both black and white were members of the same religious code (the Body of Christ), it is no surprise that after the abolition of the slave trade and later slavery itself, Europeans continued to oppress Africans. It is interesting to note that the 'Dash for Africa' in the mid-nineteenth century came soon after slavery was finally abolished in the British Empire.

The existence of racism in Britain today is testament to the continuance of the underlying Eurocentric Judaeo-Christian framework that has always caricatured Africans as 'less than' and 'the other' – i.e. not one of 'us'. So why am I still within the Christian church trying to effect an anti-racist ethic in Christian ministry? I remain a practising Christian because there is another story to be told, one that lies in the heart and mind of such luminaries as Sam Sharpe, a Baptist deacon who initiated the largest rebellion in Jamaica against slavery, in the Christmas period of 1831. For Sharpe as well as other enslaved Africans, Jesus was the Liberator who came to bring freedom to the captives (Reddie 2006). Texts such as Luke 4.16–19 or Matthew 25.31–46 became 'proof texts' that God as reflected in the life, death and resurrection of Jesus was on the side of the oppressed and also against the perpetrators of the slave trade.

Deconstructing Mission Christianity

The relationship between empire and colonialism, in which Christianity and whiteness are inextricably linked, remains the unacknowledged 'elephant in the room' in much academic theological discourse in the UK. R. S. Sugirtharajah, the doyen of postcolonial biblical hermeneutics, noted that the relationship between British Christianity and empire is one that has been suffused with a collusive sense of mutuality. Both the Christian faith and imperialism, and the regimes that connote the latter, presume themselves to be superior to the phenomenological entities they seek to usurp or supplant. Speaking with particular attention to the question of empire, Sugirtharajah writes:

> Empires are basically about technically and militarily advantaged superior 'races' ruling over inferior and backward peoples. When imperial powers invade, the conquered are not permitted to be equal to the invaders. This was true of all empires, Roman to British and American. The basic assumption of superiority is never questioned in their writings. (2003, p. 147)

The superiority of Britain is built upon a bedrock of Christian inspired notions of exceptionalism in which God has set apart the British, particularly the English, to occupy a special place in the economy of God's Kingdom. One can see an element of this in the rhetoric of Britain's greatest writer William Shakespeare, who in his play *Richard II*, written in 1595/6, a few years after the Spanish Armada of 1588, states in unambiguous tones the import of the English when thinking in terms of their sense of exceptionalism. Shakespeare writes (John of Gaunt's death-bed speech in Act 2, Scene 1):

> This royal throne of kings, this scepter'd isle,
> This earth of majesty, this seat of Mars,
> This other Eden, demi-paradise,
> This fortress built by Nature for herself
> Against infection and the hand of war,

This happy breed of men, this little world,
This precious stone set in the silver sea,
Which serves it in the office of a wall
Or as a moat defensive to a house,
Against the envy of less happier lands,
This blessed plot, this earth, this realm, this England.

The outworking of this exceptionalism was the desire to export the superiority of the British across the world. Empire and colonialism found much of its intellectual underscoring on the basis of white, Eurocentric supremacy, which marked the clear binary between notions of civilized and acceptable over and against uncivilized and transgressive. There are no prizes for guessing on which side of the divide black people found themselves.

The roots of Brexit lie in the growth of an English nationalism – which sees 'us' as different from 'them' – which really begins during the reign of Elizabeth I. The rise of English nationalism was based on notions of being different from, and better than, others. Underpinning it is a subterranean theology of election which identifies whiteness and Englishness as the defining symbol for the construct for righteousness, and as a signifier for religious acceptability. This theological underpinning of English nationalism feeds into the sense of privilege that has fashioned ideas of empire, the Church of England, and conservative politics.

Is it any wonder, then, that the trigger for the referendum vote emerged from the discontentment of English nationalism from within the Eurosceptical wing of the Conservative Party – and found subsequent support within the congregations of the Church of England?[4] The Brexit vote demonstrated the barely concealed exceptionalism and sense of entitlement of predominantly white English people. The clear xenophobia underpinning the Leave campaign reminded many of us that 'true Britishness' equals whiteness, and that those who are deemed the 'other' – be they migrants living in the UK or foreigners from Europe – are distinctly less deserving in the eyes of many white British people.

It can be argued that the romantic push for the nostalgia for the past (when Britain had the biggest empire the world has ever seen) is predicated on the intrinsic value of Britain being superior to others, often seen in terms of groups such as Britain First or groups on the political right who want to 'make Britain great again'. To quote the black British social commentator Gary Younge (2016): 'Not everyone, or even most of the people, who voted leave, were driven by racism. But the leave campaign imbued racists with a confidence they have not enjoyed for many decades, and poured arsenic into the water supply of our national conversation.' The toxicity of the hostile climate on immigration was one that has helped to create a contemporary era in which white entitlement has reasserted itself, blaming migrants and minorities for the social ills that supposedly plague the nation.

It is my contention that the vote for Brexit was very much based on the presumption of white normality, and the belief that the needs of poor, disenfranchized white people would be better served if the numbers of poor minority-ethnic people and others from outside the UK were reduced. That so many poor white people believed such blandishments can be explained, in part, by my presumption that whiteness remains a site for privileged notions of belonging, and its concomitant identity is one embedded in paradigms buttressed by superiority and entitlement.

In our current era, British people must purge ourselves of the noxious fumes of imperial grandeur that still continue to assault the senses of contemporary Britain; the church must also recover its prophetic stance as a critical voice that rejects the populist thrust of particularly white nationalism. John Hull, my doctoral supervisor and one of the most honest and critical decolonial scholars, has argued that British churches need to rediscover a prophetic edge in order to critique the inherent white nationalism that underpinned Brexit and made it more than just a seemingly transparent democratic vote to remain within or leave the EU. Hull's perceptive analysis is on the ongoing tension between the prophetic church with its roots in the Prophets of the Hebrew Scriptures and the imperially friendly church that colludes with rather than opposing oppressive, totalizing forms of white nationalism, underpinned by an unreconstructed theology of empire (see also Chapter 7 by Gerald West). Hull critiques imperial theology and recognizes that the dangerous residues of this theological framework has not gone away:

> The theology of empire has outlived the empire. The empire has gone but its theology lingers on. Much of the modern church is like the Israelites, going into exile with a royal Kingdom theology. Faith has become a remnant, far from the glories of its greatest achievements. (Hull 2014, p. 109)

Hull points to the constant flourishing of a theology of resistance to the worst excesses of imperial theology. He recognizes the radicalism of the early Methodist movement under John Wesley in the eighteenth century, the Christian socialism of Anglican social theology in the mid-nineteenth century, and the developments in contextual theologies of liberation from the global south in the latter part of the twentieth century (Hull 2014, pp. 67–160).

So where do we go from here? The major ethical and moral question, for me, has been the underlying issues and concerns that have arisen as a result of the referendum vote. We have seen a rise in racist attacks and xenophobia since 2016 and the underlying frameworks of Brexit has been the unexplored dimensions of Whiteness that have remained unresolved in the British psyche. The ways in which the Leave Campaign was able to tap into the latent fears of white people at the presence of non-white people was symptomatic of a nation that has still to come to terms with its multicultural, post-Second World War heritage. Ironically, it was the areas that were more monochrome and white that voted to Leave more than multicultural ones in many of our larger urban conurbations in England.

So, the major theological and ethical challenge for the church post the EU referendum is how can the church be in solidarity with black and minority ethnic people and vulnerable migrants, asylum seekers and refugees? Given that Brexit has emboldened groups on the political right like 'Britain First' and the English Defence League, the sharp challenge is, where is the church leadership that will face down the rise in white English nationalism?

Deconstructing Whiteness

In a context where Britain has seen a lurch to the right, the question is, how can twin pandemics give rise to a critique of the old world-order that was built on white entitlement? White English Christianity must be committed to a ruthless and fiercely argued critique of its whiteness, in a manner that accords with the existential struggling for truth that black people have been obliged and sometimes forced to undertake since the creation of modernity.[5] In critiquing whiteness, I am talking about a thorough deconstruction of toxic relationship between Christianity, Empire and notions of white British superiority (Reddie 2009, pp. 37–52).

The quest for equity, liberation and justice is one that requires the committed determined action of all peoples, irrespective of faith commitment (plus those who profess to hold no such notions). But it also requires truth-telling and a retreat from all forms of obfuscation that blind us to the structural and systematic forms of racism that continue to oppress black people and other minority ethnic people in Britain. Whether we wish to acknowledge it or not, privilege and notions of who is important, has a colour, and of course, colour is always linked invariably to asymmetrical forms of disadvantage and oppression. Similarly, systemic power and notions of belonging and what is deemed acceptable also has a colour, alongside that which is stigmatized (see Ray 2010).

The existence of racism in Britain today is testament also to the continued legacy of Mission Christianity and its contribution to the construction of whiteness. Slavery is long gone but anti-black racism outlived the institution that helped to breathe it into life. In our contemporary era, the underlying framework of blackness which still symbolically is seen as representing the problematic other, finds expression in a white police officer placing his knee on the neck of a black man and despite the plaintive pleas of 'I can't breathe' the officer remains unmoved and maintains his violent posture until this black man dies. One cannot understand the futility of this death unless you understand that this is no new phenomenon. White power has viewed black flesh as disposable for the past 500 years.

Before anyone suggests that this is a purely American phenomenon, let me recall the death of Clinton McCurbin, an African Caribbean man who died of asphyxia at the hands of the police in Wolverhampton on 20 February 1987, having been arrested for using a stolen credit card (see Flah and Hyatt 2020). The 'Black Lives Matter' movement emerged in order to counter the patently obvious fact that black lives do not matter (see Lightsey 2015).[6] This is not just a question of economics or materiality; it is also about seemingly 'ephemeral matters' like the impact on our psyche and associated questions of representation and spirituality. It has been interesting observing the concern of many white Christians for the ethical matters of law and order, governance and property, in relation to the tearing down of the Colston statue in Bristol. Black people, many of whom are descendants of enslaved peoples, have lived in that city within sight of a statue built in honour of a slave trader. Polite petitions to move these and other statues were ignored. Long before a so-called mob tore this one down, activists asked for it to be moved to a museum (where those who deliberately wanted to see it could but saving those of us who did not the ignominy of having the lives of our oppressed ancestors constantly insulted). White authority ignored our claims, because black lives and our resultant feelings do not matter. Black lives do not matter in the face of white complacency and disregard, just as our pleas for justice for Clinton McCurbin went unheeded because our feelings did not matter either.

So, I find it interesting that following the pulling down of a statue, we had the usual furrowed brow of some white Christians sharing their ethical concern for law and order and the dangers of mob rule.[7] One wonders how many of these complainants were supportive of BLM prior to its sudden resurgence since the death of George Floyd? For some respectable white Christians, their ethical concern is focused on property and not black lives disfigured by racism. Delroy Wesley Hall (2009) speaks of black people living in Britain struggling with a form of existential crucifixion. We are mired in our continued Holy Saturday following our social and collective crucifixion, with no Easter Sunday on the horizon.

So at this moment in history, I am not going to thank white people for issuing apologies, 'taking the knee', writing statements and going on marches that do not cost them anything, when we are dealing with forms of existential crucifixion that lead to us being more likely to struggle with mental ill health issues, such as schizophrenia.[8]

I am not going to 'educate' white people on how to deal with their discomfort and emotions when I and countless black people are afraid to go out of our houses lest we end up as part of the disproportionate numbers who are stopped, detained and questioned by our supposedly benign police force for violating the changeable rules on social distancing during the lockdown that see white people congregating with impunity (see Dearden 2020).

Thinking back to 1987, when I asked my white Christian colleagues and friends to support me in mounting a campaign to mark the callous killing of Clinton McCurbin, I was met with complete indifference. McCurbin's death did not resonate with them because the death of another anonymous black man was no big deal. And yet every black person knows that in and of itself, George Floyd's death is not remarkable. Systemic racism did not start with George Floyd's death nor will it end with white people wringing their hands in liberal guilt, telling us how sorry they are for the racism that blights our lives and not theirs. The bitter truth is that black lives have not mattered for a very long time and the church has long been complicit in this.

I have used the iconic toppling of the Colston statue as a microcosm for the wider Black Lives Matter movement and the indifference of some white Christians to our pleas for justice.

The frustration of the protestors that led to the toppling and disposal of the statue reminds me of the very human anger and frustration of Jesus in turning out the money changers in the temple (Matt. 21.12–17; Mark 11.15–19; Luke 19.45–48 and John 2.13–16). It seems it is all right for a 'white Jesus'[9] as depicted in Western iconography to be angry and destroy property but not unruly black people! An anti-racist ethic in Christian ministry is one that supports black lives, if white Christians are serious about seeking to be in solidarity with black people as we wrestle with the continued realities of systemic racism.

If the church is serious about Black Lives Matter, then we need more than rhetoric and well-crafted words and vacuous resolutions. The Black Lives Matter movement is considered problematic by a number of people. But the truth is, the Christianity that has been bequeathed to us by centuries of white supremacy has proved to be far more dangerous to the world than anything the Black Lives Matter movement could even conceive.

The truth is, white supremacy is not going to sue for peace anytime soon. The removal of a few statues – done voluntarily or removed by force – is not going to topple white privilege anytime time soon. My favourite white theologian working today is James Perkinson. His monumental book simply entitled *White Theology* (2004) is one of the few texts written by a white theologian that has accepted that whiteness is a problematic that cannot be imagined away by a few well-crafted prayers and a hearty rendition of Kum-bah-yah. Black lives will continue not to matter until the church has the courage to stand up to white supremacy. Perhaps the present moment is the start of a prophetic breakthrough, or it may be yet another false dawn. A great Danish theologian once observed that life is lived forward but understood backwards. Ultimately, history will judge us all.

The fact is, returning to the old normal would be the worst thing that can happen to us. One hopes that the new normal will not include Covid-19 or racism. My hope is that anti-racism and a commitment to the fundamental tenets of liberation

theologies and the preferential option for the most marginalized and the oppressed (see Chapter 14 by Zwane), especially
those in darker-skin bodies (see Balasuriya 2002), will be the
new mode of the new normal. This may be an unrealistic hope,
but what is a people, indeed any people, without hope? I am
committed to a liberative, prophetic hope.

Notes

1 For an incisive and critical interrogation of the corruption of
Christianity by notions of 'Race', which assisted in the theological construction of chattel slavery see Carter 2008.

2 For a critical rereading of the Exodus narrative, which offers an
anti-imperialist, anti-hegemonic hermeneutic see Warrior 1997.

3 For a wider discussion on the destruction consequences of the
book of Joshua, see Bridgeman 2010.

4 Research by Linda Woodhead has shown Anglicans in the Church
of England were more likely to have voted for Brexit than any other
Christian community or church in the Referendum on the European
Union. Given that the Referendum emerged from within the discontent
of the Euro-sceptical wing of the Conservative Party, it would appear
that the Church of England remains the Tory Party at prayer. For
further insights on this, see Taylor 2018.

5 Perhaps the best work that addresses issues of Whiteness and privilege in Christian theological terms is Webster 2009.

6 Lightsey examines the plural and intersectional nature of the Black
Live Matter Movement and affirms Black LGBTQI+ people, in order to
reassert the primacy of all black bodies mattering and not just respectable heteronormative, churchgoing ones.

7 This comment is reflective of the push-back of 'some' white
Christians on social media responding to the threat to law and order
and property. It is important to acknowledge the many black Christians
who have also shared their disquiet at the dangers of mob rule and
the desecration of public monuments. I am forced to acknowledge that
there are obvious dangers of untrammelled 'violent' direct action of this
sort. My comments are not an absolute endorsement of this action but
on the complicity of the authorities in the city to side with the blandishments of white supremacy that is exemplified in the maintenance of a
statue of Edward Colston in the first place.

8 For an excellent exploration of the ways in which black Christian
faith has been an essential means of dealing with the environmental
features of systemic racism that leads to disproportionate levels of mental
ill health among African Caribbean people in Britain, see Willis 2006.

9 Arguably the most comprehensive appraisal for the epistemological frameworks that have given rise to the predominance of a White
Jesus, see Kelley 2002.

References

Tissa Balasuriya, 2002, 'Liberation of the Affluent', *Black Theology: An
International Journal* 1.1: 83–113.
Valerie Bridgeman, 2010, 'Joshua', in Hugh R. Page Jr. *et al.* (eds),
*The Africana Bible: Reading Israel's Scriptures From Africa and The
African Diaspora*, pp.180–8, Minneapolis: Fortress Press.
J. Kameron Carter, 2008, *Race: A Theological Account*, New York:
Oxford University Press.
Lizzie Dearden, 2020, 'Police fining and arresting black and Asian people
disproportionately under coronavirus laws in London', *Independent*,
3 June (accessed 4.8.20: www.independent.co.uk/news/uk/crime/
police-fine-arrest-black-people-coronavirus-lockdown-london-a954
6181.html).
Kelly Brown Douglas, 2005, *What's Faith Got to Do with it?: Black
Bodies, Christian Souls*, New York: Orbis.
Oprah Flah and Rakeem Hyatt, 2020, 'Families of dead police custody
victims speak about fight for justice', *BirminghamLive*, 19 June
(accessed 25.7.20: www.birminghammail.co.uk/black-country/fam
ilies-dead-police-custody-victims-18412737).
Gretchin Gerzina, 1995, *Black London: Life Before Emancipation*,
New Brunswick, NJ: Rutgers University Press.
Delroy Hall, 2009, 'The Middle Passage as Existential Crucifixion',
Black Theology: An International Journal 7.1: 45–63.
Robert E. Hood, 1994, *Begrimed and Black: Christian Traditions on
Blacks and Blackness*, Minneapolis: Fortress Press.
Dwight N. Hopkins and Marjorie Lewis (eds), 2014, *Another World is
Possible: The Spiritualities and Religions of Global Darker Peoples*,
London and New York: Routledge.
John M. Hull, 2014, *Towards The Prophetic Church: A Study of
Christian Mission*, London: SCM Press.
Sylvester A. Johnson, 2004, *The Myth of Ham in Nineteenth-Century
American Christianity: Race, Heathens, and the People of God*, New
York: Palgrave Macmillan.
Shawn Kelley, 2002, *Racializing Jesus: Race, Ideology and the Forma-
tion of Modern Biblical Scholarship*, New York: Routledge.
Pamela Lightsey, 2015, *Our Lives Matter: A Womanist Queer Theology*,
Eugene, OR: Pickwick.

James W. Perkinson, 2004, *White Theology: Outing Supremacy in Modernity*, New York: Palgrave Macmillan.

Stephen G. Ray, 2010, 'Contending for the Cross: Black Theology and the Ghosts of Modernity', *Black Theology: An International Journal* 8.1: 53–68.

Anthony G. Reddie, 2006, *Black Theology in Transatlantic Dialogue*, New York: Palgrave Macmillan.

——, 2009, *Is God Colour Blind?: Insights from Black Theology for Christian Ministry*, London: SPCK.

R. S. Sugirtharajah, 2003, *Postcolonial Reconfigurations: An Alternative Way of Reading the Bible and Doing Theology*, London: SCM Press.

Ros Taylor, 2018, 'How Anglicans tipped the Brexit vote' (20 Sept; http://blogs.lse.ac.uk/brexit/2018/09/20/how-anglicans-tipped-the-brexit-vote/; accessed 21.9.18).

Robert Allen Warrior, 1997, 'A Native American Perspective: Canaanites, Cowboys and Indians', in R.S. Sugirtharajah (ed.), *Voices From The Margin: Interpreting the Bible in the Third World*, pp. 277–85, Maryknoll: Orbis.

Alison Webster, 2009, *You Are Mine: Reflections on Who We Are*, London: SPCK.

Lerleen Willis, 2006, 'The Pilgrim's Process: Coping with Racism through Faith', *Black Theology: An International Journal* 4.2: 210–32.

Gary Younge, 2016, 'After this vote the UK is diminished, our politics poisoned', *The Guardian*, 24 June (accessed 23.5.17: www.theguardian.com/commentisfree/2016/jun/24/eu-vote-uk-diminished-politics-poisoned-racism).

22

for everyone?

writing a new script
inviting the impossible
in a world ravaged and changed by diseases

no place safe
no bodies sacred

brothels, beaches, and bodies isolated and ignored
all closed because of uprising and upheaval on all shores

bodies shut down by inequalities
gender, race and economy exposed

attacking the body in every form
disproportionally engaged in an ill-timed assault
ravaging human bodies

redefining institutional bodies
changing the body of Christ as we know it

whose new world is emerging

transforming environments
bringing unscripted change

what to make of this disease
what is the message we hear

from this pandemic
monstrosity
what are we willing to grieve

this Covid-19 with its myriad of complexities
brings with it death of body, mind and spirit

body count soaring as the days go by
dead bodies rejected

identified as feared as they too are terrorized
missing rituals of closure
their spirits still journey on absent of funerals and mourning

whose new world are we seeking

immoral redefined
families left unprotected
criminalized, racialized and problematized
marginalized bodies charged to bear more harm and pain
justice is yet evasive on their behalf

vulnerabilities of lives ostracized by unjust social conditions
Covid-19 one more threat to lives inhabiting bodies
    demonized
what is the message brought into our midst

what is the message brought by the falling
falling bodies
falling angels

falling morals
falling empathy
falling gods

whose new world are we imagining

no safe zone to be found in the world
we continue to listen for active resistance
to the problems of life exposed

what do we want to see socially realized
knowing that the poor are expendable

indigenous, black and brown bodies do not matter
economies and empires continue to thrive

infections reveal systemic vulnerability
upending global economies
toppling multinational monstrosities

diminishing principalities and powers
there is a message from the other side

questions that come before discussions
of eternal life and the mystery of the afterlife

whose new world is fashioned by this messenger

what is the message brought into our midst
what is the message brought by this virus

exposing the powers of evil
displaying the powers of patriarchy
identifying the powers of colonialism

pointing out the powers of imperialism
uncovering the powers of transnational parasites

this pandemic teacher has much to say
bringing this advent of omen

we learn again to live as indigenous Ancestors
we learn there is more than enough
we learn everything limits and binds

we hear the message that this apocalypse
this moment is more than mere daily news

whose new world is defined by this liminality

there is potential
there are possibilities

new worlds are possible among the living
rites of passage identified to be explored

relocating ourselves
in symbols of deeper meeting
we emerge changed moved to deeper spiritual living

whose new world is problematized
navigating away from uncaring societies

whose new world is redefined
in this battle for sustainable economies

whose new world is transformed
in seeking to relieve suffering in the streets

whose new world is emerging

Karen Georgia A. Thompson
13:55
31 October 2020
Olmsted Township, OH

PART 4

In Protest

23

An Infinite Present:
Theology as Resistance
Amid Pandemics

L. JULIANA CLAASSENS

Covid did more than succeed in bringing all of us to an individual and collective halt. All around the world, trapped in our respective situations of lockdown, we were faced with the formidable task of trying to make sense of this all-consuming, turning-the-world-upside-down pandemic. One day, while flipping through Flipboard, an article with the title 'Without Future Plans, We're Living in an "Infinite Present"' caught my attention. This phrase, 'An Infinite Present', went viral after being shared by American journalist Helen Rosner to capture her frustration of living in a Covid world with 'no future plans, no anticipation of travel or shows or events or celebrations. It's an endless today, never tomorrow' (Moss 2020). The sentiment behind this phrase resonated with many. Including me.

One of Covid's side effects was that we have found ourselves caught in the unending cycle lamented also in Ecclesiastes 1. All rivers run into the sea, but the sea never is full (v. 7). The sun rises; the sun sets, only to rise once more tomorrow (v. 5). Round and round the wind blows, only to return to where it started (v. 6). Zoom meetings. Homeschooling. Teaching online. Walk around the yard. Stress baking. Watch Netflix. Repeat. For much of 2020, we were stuck in an infinite present with no vision of tomorrow and frustrated by our inability to plan for a future.

But perhaps even grimmer, we have been hurled into a world

of Jeremiah where we were no longer allowed to celebrate weddings. Or to enter a house of mourning where we may engage in rituals of lament and comfort of the bereaved (Jer 16.5–9). In terms of the inventory of seasons in Ecclesiastes 3, it seems as if for most of 2020, all around the world, there was no longer a time for mourning and weeping, or laughter and dance.

And yet, in both Ecclesiastes and the book of Jeremiah, one finds, amid a time of uncertainty and duress, traces of a theological response that refuses to accept the new normal. This essay will explore the possibilities of doing theology amid Covid in terms of the theme of resistance by fostering a creative conversation between the Sage of the book of Ecclesiastes and the Prophet of the book of Jeremiah. Each of these two biblical figures has had to endure their share of trauma and travail – the prophet Jeremiah, in particular, speaking about the sons and daughters of Judah who 'shall die of deadly diseases' (Jer. 16.4). However, in both these biblical books, one finds valuable resources to help us consider how to continue to live amid the infinite present wrought upon us by this current pandemic.

The Present Infinitive in the Book of Jeremiah

It is during this time of the Infinite Present, which had become our new normal during Covid, that I was drawn to the notion of what I describe as the 'Present Infinitive'[1] in the book of Jeremiah. Born out of the devastation of imperial attacks, followed by famine and deadly pestilence sweeping through the land, this book reflects the efforts of the traumatized prophet, representing a traumatized community to make sense of a calamity that far exceeds what we have been facing in 2020 that forever will be known as The Year of the Plague.[2]

However, amid the honest and raw expression of a national crisis, which affected individuals in very concrete ways, one finds in the middle of Jeremiah a couple of chapters in which the prophet is doing theology in a time of disaster in a way that defies his current circumstances. In a context in which people have been uprooted and torn down (cf. the recurring refrain 'to pluck up and to pull down, to destroy and to overthrow, to build and to plant', Jer. 1.10), the intended audience in Jeremiah is told to resist their situation by engaging in a number of basic, everyday activities, presented in the form of the present infinitive (Claassens 2018, pp. 695–6).

The so-called Little Book of Comfort starts with Jeremiah (in Jer. 29) writing a letter to the exiles in Babylon. In this reality therapy of sorts, Jeremiah is telling the exiles to accept the situation in which they find themselves; to settle in for the long haul for this exile is going to last a while – for most of the addressees, it will last for their entire lifetime!

But acceptance does not mean passivity. The prophet's exhortation to engage in present active infinitive verbs: to build, to plant, to build a family, to engage in diplomacy, represent tenacity, or one could say, persistence, insistence, determination, stubbornness, resolve – all synonyms that reflect the call not to give up; to continue to live amid the misery that had befallen them.[3]

All these activities require hard work – daily, laborious activity that in the end yield fruit and give life. In a context in which houses were torn down, the exiles are called upon to build again; to plant vineyards as a direct response to the enemy's scorched earth policy. It is these acts of engaging in everyday actions, mundane and special occasions alike which are at the root of theology as resistance in the book of Jeremiah.[4] In particular, the call to have your children marry, and your children's children, is significant given the absence of weddings and the prohibition of mourning in Jeremiah 16. Indeed, human beings crave ritual and will find ways to celebrate unions and mourn loss – even amid Exile.

But the prophet is not merely doling out advice to others. In chapter 32, Jeremiah actively lives his own advice where he finds himself, confined in the house of the prison warden – in a sense doubly trapped as the city of Jerusalem is beleaguered by the Babylonians. In this situation which we may now describe as 'lockdown', the prophet himself is not passive but rather acts in a way that mirrors the advice he has been imparting on to the exiles in Jeremiah 29.

In a profound act of resistance, the prophet in Jeremiah 32 engages in a counter-intuitive real-estate transaction, which serves as a primary means of expressing Jeremiah's agency and subjectivity, as the prophet is shown to buy a portion of land at the request of his cousin who had come to visit him in prison. To buy land implies to have land on which one is able to conduct the life-giving activities in which Jeremiah had implored the exiles in Jeremiah 29 to engage: building homes, planting vineyards, building families.[5]

This act of buying property, captured in great detail with Jeremiah being the subject of at least eight actions that involve transferring funds and signing the deed in the presence of witnesses (vv. 9–12), signifies normality, a life from before the war which also forms the basis of the hope for a future when some semblance of normality will return. As the prophet himself explains his actions in terms of the divine promise that one day 'houses and fields and vineyards shall again be bought in this land' (Jer. 32.15).[6]

What ties these two chapters together is the vision of the people returning to the land that once more employs images of building and planting to symbolize hope for the future. Offering a sharp contrast with the exiles' experience of being uprooted and broken down, we read in Jer. 31.4–5 the following hopeful promise:

> Again I will build you, and you shall be built, O virgin Israel!
> Again you shall take your tambourines, and go forth in the dance of the merrymakers
> Again you shall plant vineyards on the mountains of Samaria; the planters shall plant, and shall enjoy the fruit. (NRSV)

It is this belief that the Builder/Planter God will once more build and plant Judah, hence causing her people to flourish, which makes action in the present possible as evident in both Jeremiah's letter to the exiles as well as his own life amid lockdown.

Perhaps I could resonate with Jeremiah's message of building, planting, and buying a place in which such life-giving activities might ensue, because, during those days of the extremely hard lockdown my country South Africa was living through, present infinite actions of building, planting and building relationships served as a way to offer resistance against the tyranny of the Infinite Present.

For five very long weeks, we were not allowed to go out of the house except to go buy food or seek medical attention. Even exercise was prohibited, together with alcohol and cigarettes of all things! During this time, people began to engage in small life-giving actions that sought to cling to beauty and joy. Mundane acts such as growing herbs in one's windowsill, and cultivating a sourdough starter, which may be said to be a contemporary form of keeping a plant alive, became part of the daily battle against the darkness and gloom associated with this very strange year.

I was in particular drawn to Jeremiah's act of buying land, as, quite ironically, my husband and I engaged in the counterintuitive act of buying a house ourselves, and miraculously selling the home where we for the past 14 years had built a family and a life, during lockdown! Suddenly this Year of the Plague was filled with new meaning, as we transitioned from one life to another, practising the art of letting go and of travelling lighter, as we tried to figure out which of our furniture and possessions would fit into this new space. The act of buying a house in which we could continue to build a life and plant a vineyard was to all of us a profound expression of subjectivity, of resistance in a world, turned upside down, which was becoming gloomier by the day.

And yet, while writing this reflection, and particular in my engagement with colleagues from around the world, brought together in a truly global endeavour as we jointly reflected on the question of 'Doing Theology in the New Normal', I realized all too well my situation of privilege. Of having a house with a garden. Of having a space in which to shelter during lockdown. I was keenly aware that my situation is *not* the reality of so many others in my country and on my continent. Indeed, for the many South Africans living in subhuman conditions, the destitute and the homeless, lockdown and social isolation have been all but impossible. Or impossibly difficult.

I also saw with new eyes Jeremiah's privilege – despite being locked up, having the means to buy property from a family member who owned land. I considered anew the privilege of the exiles in Babylon who were the elite of Jerusalem: the artisans, the skilled workers, those with means.[7] Even though brought there in shackles, the exiles likely continued to live relatively privileged lives in their new country as evident also in Jeremiah's imperative to build houses, plant vineyards and hold weddings. Both Jeremiah's situation, as well as that of the exiles in Babylon, thus offered a sharp contrast to that of the Poor of the Land, and the many refugees in Jeremiah's day who had no choice but to seek sanctuary in Egypt.[8]

What's more, the weekend just before we moved, while busy painting a sad bathroom cupboard the most beautiful bluish-grayish colour; while hanging curtains and imagining where our furniture should go, messages from all sides started pouring in. One of my first MTh students, who since graduation had served as a pastor in a small town about four hours from Cape Town, had succumbed to this dreaded disease.

I was struck by the surreality of it all. While I was painting, planting and building, actively celebrating life, beauty and joy amid the pandemic, another gentle soul succumbed to Covid-19. The Year of the Plague has indeed turned into The Year of Anguish for many as a recent newspaper article aptly describes 2020 (Baertlein and Moore 2020). This article estimates that the 200,000+ Americans (at the time of publication of this book, more than 508,000) who up till then had died due to Covid-19, each left nine individuals bereaved: significant others, mothers, fathers, siblings, children, not to speak of the extended family, friends and the community as a whole who mourn the demise of every covid victim.

I suddenly wondered whether Jeremiah's message in the Little Book of Comfort of engaging in the Present Infinitive in order to counter the Infinite Present still holds water. It is at this point that I turned to a second biblical text from the pen of the Sage Qohelet, whose counter-voice in the book of Ecclesiastes emerges out of a profound sense of disillusionment as expressed in the refrain: 'Everything is *hebel*.' According to Leong Seow, *hebel* may 'refer[s] to anything that is superficial, ephemeral, insubstantial, incomprehensible, enigmatic, inconsistent, or contradictory'. As Seow argues (1997, p. 47):

> Something that is *hebel* cannot be grasped or controlled. It may refer to something that one encounters or experiences for only a moment, but it cannot be grasped neither physically nor intellectually.

Life itself is fleeting. Gone before you know it. Reminiscent of the metaphors employed in Ecclesiastes 12's poem to old age of a clay pitcher being broken, and a silver chord being snapped (v. 6), Covid-19's most tragic legacy is that chords were being cut and pitchers shattered for many people around the world, long before it was their time for dying (Eccles. 3.2). Coupled with the so very realistic description of the Infinite Present cited in the beginning of the essay, it seemed that the sentiments reflected in the book of Ecclesiastes might be a most fitting conversation partner for doing theology in a time of pandemics when Jeremiah's poignant words of resistance fall short.

The Internal Infinite in the Book of Ecclesiastes

The book of Ecclesiastes is profoundly aware of the ephemerality of everything, including life itself, the vexation of the work, and the pursuit of happiness that causes so much mental anguish and suffering to human beings. Ecclesiastes 2.23 expresses this point: 'For all their days are full of pain, and their work is a vexation; even at night their minds do not rest.' And yet despite the realization of the 'infinitely flowing stream that never fills', 'an eternally recurring wind', and a perpetually rising and setting sun (Eccles. 1.5–7), Davis Hankins (2015) argues that human beings' actions are steered by what he describes as the 'Internal Infinite' (p. 48). Based on the reference in Ecclesiastes 3.11 that God had placed a sense of the eternal or the infinity (עוֹלָם) in people's minds, while at the same time being unable to find out God's intentions, Hankins argues that desire, which he describes as the urge 'to seek satisfaction in the objective world', is an integral part of the human condition (p. 52). In terms of the 'Internal Infinite', human beings are inclined to, in a fundamental expression of human subjectivity, continuously repeat this never-ending pursuit of fulfilling one's desires (p. 48). Or as Carey Walsh so poignantly expresses this link between desire and being alive: 'Ecclesiastes grasped that common wisdom when he equated the end of desire with the very end of life. It is all over when one stops desiring (Eccles. 12.5), for life is defined by desires' (p. 35).

It is precisely amid the reality of the 'Infinite Present' that inform human beings' incessant pursuit for meaning, work and pleasure, that Qohelet in a manifestation of the 'Internal Infinite' exhorts his readers to immerse themselves fully in their immediate contexts by embracing their passions. As Hankins (p. 57) argues:

> Qohelet's commendations do not merely endorse 'the glory of the ordinary', they summon readers to examine what they ordinarily find glorious. Qohelet does not teach us to value the mundane; he beckons us to break through life's mundanity in the name of what we value.

In his exploration of the 'Internal Infinite', Hankins engages with the work of Gilles Deleuze who speaks of the way a subject 'contents itself with what a milieu gives it or leaves to it' (1986, p. 44). According to Deleuze, '[t]his contentment is not resignation, but a great joy in which the impulse rediscovers its power of choice, since it is, at the deepest level, the desire to change milieu, to seek a new milieu to explore, to dislocate, enjoying all the more what this milieu offers, however low, repulsive, or disgusting it may be' (p. 129).

In conversation with Deleuze, Hankins argues that Qohelet is using a series of 'concrete, quotidian activities' (eating,

drinking, loving one's significant other) to bring about 'affective experiences and [new] social relations' (p. 57). Hankins
writes (p. 57):

> Affective experiences that do not simply follow the dictates
> of life's material conditions are made possible because human
> being is folded by an internal infinite (עולם) that keeps sub-
> jects at least minimally distanced from the substance of their
> lives. At its most elementary, then, enjoyment in Ecclesiastes
> is an affective register of this point of contact-and-disjunction
> between subjects and the pleasurable and unpleasurable
> conditions of their lives.

Qohelet, even though experiencing some sense of alienation
that 'requires one to assess what one values and the quality of
one's attachments to others and to the world' calls on people
not to disengage from their concrete reality, but rather to focus
on what attaches us to it. It is by bringing this 'reflective … distance to consciousness', that Qohelet is exhorting his 'readers
to live into it' (Hankins, p. 58). One thus could perhaps say
that it is within the Infinite Present, compelled by the Internal
Infinite, that readers are to engage in the Present Infinitive.

Conclusion

This time of Covid has taught us things that we perhaps would
have preferred not to learn. We have come to grasp the finitude
and fragility of human life – of how a tiny airborne virus
shaped like a crown could play havoc with people's lives that
are not as invincible as they thought they were. What's more,
much around us has been uprooted and demolished as this
pandemic has caused damage that will take many years, maybe
even decades, to undo. Unemployment has skyrocketed; restaurants, small businesses, tourism, and the wine industry in
my country have been brought to their knees, not to talk of air
travel that will greatly affect our ability to travel and connect
as we have done before.

On a personal level, we continue to deal with the effects of Covid on our individual and collective well-being. The inhumane demands of online teaching and homeschooling; the fraying and dissolution of social bonds in a time of social distancing that in many instances have turned into social alienation; the loss of freedom of movement; of agency; of being able to plan for the future – all things we have taken for granted in a pre-covid world.

And in terms of our global community, as was so evident during the eDare webinars spanning time zones and continents and countries, before vaccines arrive and are distributed not just to the wealthy and the powerful but to all corners of our world, we will be stuck in an Infinite Present for a while. For as we in the Southern Hemisphere come up for breath and enjoy being in the outdoors in spring and summer, our brothers and sisters in the Northern Hemisphere are dealing with the devastating effects of a second and third wave that seem to be even more deadly than the first. And when spring has sprung in the Northern Hemisphere, we down south will enter yet another winter season with the increased threat of community spread. The virus that started in Asia and circled the globe has connected us in ways that we will be processing for a while.

In this reflection of doing theology in a time of pandemics (for sadly Covid-19 may not be a once in a 100-year occurrence!), we connected in a type of 'virtual encounter' – what Griselda Pollock describes as an act of 'transsubjective border-linking' (2010, p. 860; 2009, p. 48) – with the Sage Qohelet and the Weeping Prophet Jeremiah. As we in an act of solidarity reached across the centuries to a world far away, we realized that we are not alone after all.

We learned from Qohelet a realistic acknowledgement of the reality of the Infinite Present that includes also the recognition of the finitude and fragility of life. Nevertheless, this keen awareness of the fleetingness of human life and all of human endeavours is taken up in the eternity of God. We saw how, despite it all, Qohelet could emerge as the Preacher of Joy (Whybray 1982, pp. 87–94).[9] His recurring exhortation to eat, to drink, and to find joy in one's work as well as in one's relationships with significant others (Eccles. 9.7–9; cf. also 2.24; 3.12; 3.22; 5.17; 8.15; 11.7—12.1) matches that of Jeremiah, whose advice of engaging in present infinite activities has been so helpful to myself and to others in maintaining some sense of self amid the insanity of this past year.

Echoing Jeremiah's refrain concerning building and planting, Ecclesiastes resigns himself in the fact that there will be times for building and planting, while also acknowledging those times in which breaking down and uprooting enviably will transpire.[10] And in contrast to the book of Jeremiah in which we have encountered a failure to mourn, perhaps brought about by the debilitating effects of trauma that completely left people numb and unable to express their emotions in the wake of the devastation the community had lived through (O'Connor, pp. 20–5), Ecclesiastes acknowledges the inevitable reality of dying and weeping and mourning. In this time of Covid, we have become keenly aware of the loss of life, the loss of income, the loss of security and well-being, with those who already have been vulnerable due to factors associated with systemic injustice, being affected in particularly painful ways. As will be evident in Chapter 24 byTat-siong Benny Liew, and Chapter 25 by Dorothea Erbele-Küster, this is a reality that desperately ought to be defied as mourning and protest turns into 'good grief' and 'rage [becomes] beautiful'.

Thinking of Ecclesiastes together with Jeremiah, though, perhaps we are reminded that in times of uprootedness, we plant. In times when everything is torn down, we build. And in times of pandemics, we lament for what and for whom is lost, all the while looking for ways to desperately cling to life and love and beauty and joy. What else is there for us humans to do?

Notes

1 Note the following definition on the Linguapress English Grammar website: 'The active present infinitive can be defined as "the basic or root form of a verb" and in English typically takes the form, it can take two forms, with or without the particle to. For example: "live" or "to

live, love or to love, think or to think"' (https://linguapress.com/gram mar/infinitive.htm).

2 Cf. Louis Stulman's notion of the biblical prophets, and in particular Jeremiah as 'meaning-making literature for communities under siege' in which traumatized individuals and groups were helped in making sense of their suffering (2014, p. 185; cf. O'Connor 2011, p. 2).

3 In terms of trauma hermeneutics, this emphasis on the resumption of everyday activities may be understand as a good sign and an expression that the survivors are able to start reconnecting with life again (cf. Claassens 2014, p. 73; Hermann 1997, p. 3).

4 Stulman describes Jeremiah's message to the exiles in the following way: 'put down roots, affirm the bonds of family, and work toward peace and community building in [your] own neighborhoods' (2005, p. 256). However, I have also argued elsewhere how Jeremiah's letter to the exile conceivably could also be read as 'imperialistic propaganda' in which the prophet implores the exiles in Babylon 'to submit to imperial rule and to seek the welfare of the Empire's capital city'. I argued, based on the work by Christl Maier (2013, pp. 143–7), that '[t]his principle is closely aligned with the people's own survival as individuals and a people as a whole and hence could be viewed as the best pragmatic solution in an impossible situation' (Claassens 2019, p. 2).

5 Cf. Steed Vernon Davidson's argument with regard to the way in which Jeremiah's actions is a way in which people in a situation of subjugation are able to make for themselves in bell hooks' terms, a 'homeplace', i.e., a space in which they may exert their own sense of self and reclaim a measure of agency under imperial rule (Davidson, p. 87; cf. also Claassens 2014, pp. 78–9).

6 This text in Jeremiah 31.12 can be read in different ways. On the one hand, as Davidson has argued, this emphasis on the return of economic prosperity may serve as 'a suitable propaganda for the promotion of the territory as a viable place in which to live' (p. 115). Davidson rightly points out that the colonizers would after all benefit from a thriving economy that would result in regular tribute payments. However, the longing for a piece of land away from the imperial gaze also may be an important element in preserving the people's subjectivity and autonomy and hence serves as a counter-imperial discourse (Davidson, p. 105).

7 Cf. my essay on the variety of perspectives in the book of Jeremiah of how to survive the Babylonian invasion and its aftermath. In terms of the section headings in the essay, people were 'forced to go' (exiles), 'bound to stay' (inciles), and 'compelled to flee' (refugees) which invite important perspectives in our current context of global migration in which there is no good option (Claassens 2019).

8 Madipoane Masenya (2010) emphasizes the role of privilege in the case of the exiles who end up as the primary writers and editors of

the biblical literature that emerged in the shadow of the Empire. She argues, 'not only is the Babylonian Exile glorified, but it also occupies more space in the book. Little, if any space is given to those who went to exile in Egypt. Similarly, those who remained in Judah are portrayed negatively, the poorest of the land (40.7)' (p. 154).

9 Cf. also William Brown's argument that 'the gift of enjoyment' is inextricably connected with 'the futility of gain' and constitutes 'two sides of the same coin of Qohelet's message' (2000, p. 127).

10 This is also one of the reasons why Seow (p. 160) prefers to translate the preposition ל and the infinitive construct in Ecclesiastes 3 as a gerund, i.e., 'a time for birthing, a time for dying, a time for killing'. Seow is of the opinion that these times are just happening in life and humans can merely respond to such times. As Whybray formulates this view: 'they are the stuff of human life, and must be recognized and reckoned with' (1991, p. 481).

11 References to the titles of the essays by Tat-siong Benny Liew, 'Good Grief: Mourning as Remembrance and Protest', and Dorothea Erbele-Küster, 'Today I let my rage be beautiful': Poet(h)ical Responses to Crisis in the New Normal' that follow after my essay.

References

Lisa Baertlein and Angela Moore, 2020, 'As U.S. Covid-19 deaths near 200,000, a nation grapples with grief', *Reuters*, 21 September (accessed 12.1020: https://in.reuters.com/article/health-coronavirus-usa-grief/as-u-s-Covid-19-deaths-near-200000-a-nation-grapples-with-grief-idINKCN26C13A).

William P. Brown, 2000, *Ecclesiastes*, Louisville, KY: John Knox Press.

L. Juliana Claassens, 2014, 'The Rhetorical Function of the Woman in Labor metaphor in Jeremiah 30–31: Trauma, Gender and Postcolonial Perspectives', *Journal of Theology for Southern Africa* 150: 67–84.

———, 2018, 'Jeremiah', in J. Aguilar, R. Clifford, D. Harrington (eds), *The Paulist Bible Commentary*, pp. 666–713, New York: Paulist.

———, 2019, 'Going Home? Exiles, Inciles and Refugees in the Book of Jeremiah', *HTS Teologiese Studies/Theological Studies* 75.3, a5149 (accessed 22.3.21: https://hts.org.za/index.php/hts/article/view/5149/12305).

Steed Vernyl Davidson, 2011, *Empire and Exile: Postcolonial Readings of the Book of Jeremiah*, Bloomsbury T&T Clark, London.

Gilles Deleuze, 1986, *Cinema 1: The Movement-Image*, Hugh Tomlinson and Barbara Habberjam (trans.), Minneapolis: University Minnesota Press.

Davis Hankins, 2015, 'The Internal Infinite: Deleuze, Subjectivity, and Moral Agency in Ecclesiastes', *JSOT* 40.1: 43–59.

Judith Hermann, 1997, *Trauma and Recovery: The Aftermath of Violence – From Domestic Abuse to Political Terror*, New York: Basic Books.

Christl M. Maier, 2013, 'God's Cruelty and Jeremiah's Treason: Jeremiah 21:1–10 in Postcolonial Perspective', in Christl M. Maier and Carolyn J. Sharp (eds), *Prophecy and Power: Jeremiah in Feminist and Postcolonial Perspective*, pp. 132–49, London: Bloomsbury T&T Clark.

Madipoane Masenaya (Ngwan'A Mphahlele), 2010, 'Jeremiah', in Hugh R. Page *et al.* (eds), *The Africana Bible: Reading Israel's Scriptures From Africa and the African Diaspora*, pp. 147–56, Minneapolis: Augsburg-Fortress Press.

Rachel Moss, 2020, 'Without Future Plans, We're Living In An "Infinite Present,"' *HuffPost* (May 17 (accessed 22.3.21: www.huffingtonpost.co.uk/entry/without-future-plans-were-living-in-an-infinite-present-heres-why-thats-hard_uk_5ebe6f1bc5b6500cdf66b468).

Kathleen M. O'Connor, 2011, *Jeremiah: Pain and Promise*, Minneapolis, MN: Fortress.

Griselda Pollock, 2009, 'Art/Trauma/Representation', *Parallax*, 15.1: 42–48 (http://dx.doi.org/10.1080/13534640802604372).

——, 2010, 'Aesthetic Wit(h)nessing in the Era of Trauma', *EurAmerica* 40.4: 829–86.

Choon-Leong Seow, 1997, *Ecclesiastes: A New Translation with Introduction and Commentary*, The Anchor Bible 18B, New York: Doubleday.

Louis Stulman, 2005, *Jeremiah. Abingdon Old Testament Commentaries*, Nashville, TN: Abingdon.

——, 2014, 'Reading the Bible through the Lens of Trauma and Art', in Eve-Marie Becker, Jan Dochhorn and Else Holt (eds), *Trauma and Traumatization in Individual and Collective Dimensions: Insights from Biblical Studies and Beyond*, pp. 177–92, Göttingen: Vandenhoeck and Ruprecht.

Carey E. Walsh, 2000, *Exquisite Desire: Religion, the Erotic, and the Song of Songs*, Minneapolis: Fortress Press.

Roger N. Whybray, 1982, 'Qoheleth, Preacher of Joy', *JSOT* 23: 87–98.

——, 1991, '"A Time to be Born and a Time to Die," Some Observations on Ecclesiastes 3:2–8', in M. Mori *et al.* (eds), *Near Eastern Studies*, dedicated to H. I. H. Prince Takahito Mikasa, pp. 469–83, Wiesbaden: Harrassowitz.

24

Good Grief: Mourning as Remembrance and Protest

TAT-SIONG BENNY LIEW

Most would agree, I think, that we 'live in disturbing times, mixed-up times, troubling and turbid times' (Haraway 2016, p. 1), with widespread deaths and feelings of dread. Starting with the bushfires in Australia, the year 2020 was difficult for many around the world. Things were made even more difficult with the incompetent but impetuous and imperialistic Trumpian regime of the United States of America. Trump's lassitude towards the Covid crisis laid bare that he valued the economy over humanity, while his response to the protests for black lives against police brutality demonstrated that he would not hesitate to use the military and religion to safeguard and (r)e(i)nforce his business-first policies. In this time of many deaths and much dread, with many feeling what W. E. B. DuBois calls 'not hopelessness but unhopeful' (DuBois 1907, p. 209), this chapter argues that our continual grief and mourning are much more than psychologically cathartic, but can function as a productive force to insist that things could have been different.

Politics of Grief

Anne Anlin Cheng (2001) uses melancholy to describe the ongoing grief that cannot be addressed or resolved through grievance, especially in the experience of minoritized people in the United States of America, since an Asian American cannot, in the words of Sara Ahmed, 'approximate an ideal [of white-ness] that one has already failed' (Ahmed 2014, p. 150), not to mention the loss of 'Asianness, home, and language' (Eng and Han 2000, p. 667). Cathy Park Hong also talks about how minoritized persons develop what she calls 'minor feelings', which include 'paranoia, shame, irritation, and melancholy', especially when an 'American optimism' that contradicts their racialized reality is being imposed on them (Hong 2020, p. 55). In addition to articulating melancholia 'as a uniquely-suited means through which to explore racism' (Kaplan 2007, p. 514) in terms of what Raymond Williams (1975) calls a collective 'structure of feeling', Cheng argues that this melancholy or ongoing grief can serve as a productive basis for constructing both an identity for minoritized communities and a politics of resistance. In the last couple of decades, there has been much scholarly investigation of melancholy or unfinished grief as enabling a 'politics of mourning' (Eng and Kazanjian, 2003b, subtitle) which takes place 'under conditions in which history, and the narrative coherence and direction it once promised, has been shattered' (Butler 2003, p. 471). As a politics that minoritized people perform rather than repress their disappointments and pain, unresolved mourning keeps alive and renders memorable the 'waste' and 'excess' that do not fit a plot of progress (Winters 2016, p. 101) but unsettle us with the need to re-cognize realities that we would rather forget or deflect onto other nations or peoples. By providing reminders of 'duress' (Stoler 2016) and preserving remainders of 'debris' (Stoler 2013), unfinished griefs have the potential to haunt us into a 'sociological imagination' of 'what could have been' (Gordon 1997; 2011). Mourning past wounds and losses, in other words, can help us think of or help make us aware of unrealized dreams, missed opportunities, foreclosed alternatives, or what normative politics see as impossible or impractical (cf. Eng and Kazanjian 2003a, pp. 4–5). Perform-ing a politics of mourning, then, 'functions as an episteme, a way of knowing' (Taylor 2003, p. xvi). As Yên Lê Espiritu writes, '[W]e [must] become tellers of *ghost stories*, ... pay attention to what ... history has rendered ghostly, and ... write into being the seething presence of the things that appear to be not there' (2014, p. 23).

Referencing not only Freud's work on mourning and melancholia but also Butler's insight that lives have to be recognized if their loss is to be grievable (Butler 1997; cf. Butler 2004), Ahmed suggests that a politics of grief, by publicly declaring that what others view as ungrievable losses are *not only missing but also missed* (Ahmed 2014, p. 157, emphasis original), helps to keep pushing the question about whose lives count and what losses are grievable (Ahmed, pp. 155–61). Instead of seeing both the griever and the grieved as victims, a politics of grief actually affirms both as subjects and, hence, differs from a wallowing in grief or what Edward W. Said calls a 'politics of blame' (Said 1993, p. 18). By refusing to 'let go', an enduring politics of grief is, for Ahmed, not only ethical to keep the dead alive rather than to 'kill [them] again' through a failure to remember but also enabling for both the grievers and the lost object to form new attachments (Ahmed, pp. 159, 187).

There is another aspect of Ahmed's work that is important for my purposes. Presenting emotion as providing 'a script' that generates affects, Ahmed compares affect to a speech act that is addressed to someone (Ahmed, pp. 12, 177, 215). Ahmed's book on emotions actually focuses on reading texts by talking about 'the emotionality of texts' (Ahmed, p. 12). For her, texts are not 'repositories of feelings and emotions' (Cvetkovich 2003, p. 7) so emotions are not properties contained 'in' texts (Ahmed, pp. 14, 19 n.22). They are rather 'objects of emotion' that can be circulated to generate affective effects, because emotions 'work by working through signs', including language and literary texts (Ahmed, p. 191) – so for me, by extension, the Bible.

John's Gospel of Grief

In her book titled *Passed On*, Holloway (2003) discusses how African American mourning and burial practices in response to black people's vulnerability to untimely deaths in the United

States play a crucial part in the construction of black identities. In that sense, those who have passed on may still have insights to pass on to those who are alive. Similarly, the Gospel writer was passing on a story of (a resurrected) Jesus to readers (including us now) after Jesus had already passed on by Roman execution. John's Gospel can therefore be read as a form of recalcitrant memory or even a 'militant refusal to allow certain objects to disappear into oblivion' (Eng and Han 2000, p. 695) by a colonized people. The description of bereavement remains in a politics of grief and mourning as 'register[ing] a tension between loss and survival, absence and presence' (Winters 2016, p. 50) seems to be perfectly applicable to John's Jesus.

Let's face it, John's Gospel gives us a picture of the world that is far from rosy. Grief is a story, and John's Gospel is in many ways a story of grief. Its protagonist, Jesus, is presented as the slaughtered Passover lamb (1.29, 36; 19.13–42). Painting a world full of hatred and death (Liew 2016b), John's narrative haunts us with its memory or its story of a dead-but-resurrected and present-yet-absent Jesus.

John's Jesus makes several ghostly appearances after his crucifixion and resurrection. After appearing to Mary Magdalene (20.11–18), he appears to his disciples in two consecutive weeks (20.19–29). The better-known episode between these two weekly appearances by John's Jesus is undoubtedly the second one, thanks to the 'doubting Thomas' tradition (20.24–29). John's Jesus comes across here, first of all, like a phantom, as he can obviously go through shut doors or solid walls (20.26). Then we learn, because of Thomas's expressed desire or need to verify Jesus's identity by checking out his wounds (20.24–25), that John's resurrected Jesus is actually a spectre with substance – or he has suddenly shifted from magical to material.

Because of the repeated references to faith (20.27–29) and the explicit contrast with doubt made by John's Jesus – 'Do not doubt but believe' (20.27, NRSV) – in this episode, the problem of doubt becomes for some the dominant meaning of these verses, with poor Thomas becoming the symbol of scepticism. Regardless of how one may want to read Thomas, there is much more to think about besides the question of doubt. Notice that John's Jesus himself, without anyone asking, *volunteers* to show his hands and his side – so supposedly the proofs of his crucifixion and resurrection – to his disciples when he appeared to them a week earlier (20.20). In other words, he shows his other disciples basically what he shows Thomas, so it is not only because of Thomas's doubt that he shows his hands and his side.

In a chapter within her recent book, Candida Moss focuses on this Johannine episode and observes that there has not been adequate scholarly attention to the resurrected body of John's Jesus (Moss, pp. 22–40). Citing sources such as Homer, Plato, Virgil, Galen and Christian readers from late antiquity, Moss argues that reading John's Jesus as showing his disciples and Thomas his crucifixion scars rather than his crucifixion wounds would make more sense, especially given the assumption that John has Jesus showing his hands and side to authenticate that he is indeed the crucified Jesus and/or to confirm that he is not an apparition. This is so for Moss because scars (1) were often used in the Greco-Roman world to identify people; and (2) should have developed on the *physical* body of John's Jesus more than a week after his crucifixion if he was not a ghost. Having said that, Moss admits that the wound that John's Jesus suffered on his side might have taken longer to heal, and that there is an existing and influential tradition that reads Jesus' hands as covered with scars but his side as an open wound. Given the ambiguity of the Greek, which Moss also acknowledges (pp. 28–9), I tend to think that we can go with either 'scars' or 'wounds', or both, especially since they bring different nuances to these back-to-back appearances by John's Jesus.

Reading and Feeling the Scars

In making her argument, Moss cites Philo's use of the Greek word *typoi* 'to describe the impressions left by old wounds' (Moss, p. 29). Interestingly, Ahmed reminds her readers that we need to '*remember the press in an impression*' (Ahmed,

p. 6, emphasis original) as she discusses how emotion or affect 'sticks' or 'leaves its mark or trace' on 'objects of emotion' with which we come into contact and, in the process, presses upon these objects to form or 'intensify' surface and boundary by potentially 'align[ing] bodily and social space' (Ahmed, pp. 6, 11, 15, 70). Sharing Moss's concerns regarding rushing to healing, Ahmed is wary of state narratives of 'moving on'; Ahmed asks us to 'rethink our relation to scars, including emotional and physical scars', because their 'lumpy' covering *always exposes the injury* of the past in the present (Ahmed, pp. 201–2, emphasis original). Scars can remind us that 'recovering from injustice cannot be about covering over the injuries', and that 'justice involves feelings' (Ahmed, p. 202).

We know the injuries that John's Jesus suffers are physical. Not only was he physically abused and crucified, but his dead body was also jabbed and gashed by a spear (18.22; 19.1, 3, 16–18, 34). Moreover, his injuries were also emotional. He was derided before his crucifixion (19.2–3), and he worries about his mother and his disciples when he is hanging on the cross, so he asks them to care for each other (20.26–27). We also sense his emotional struggles when he, anticipating his arrest and his 'hour', openly tells his disciples that his 'soul is troubled' (12.27). His farewell discourse (John 14—17) is also full of anguish, as he tries to comfort, warn, and teach his disciples as well as pray to his 'Father' all at the same time. He talks about his disciples being troubled (14.1), filled with sorrow (16.6), and how they will weep, mourn and have pain (16.20–22).

John's Gospel makes it clear that the physical and emotional scars on the body of John's Jesus, if we choose to side with Moss's preference, are results of injustice. Caiaphas is willing to sacrifice John's Jesus to prevent a military attack by the Romans (11.45–53; 18.14). While Annas and his police have no response when John's Jesus challenges them to name his transgression (18.19–24), Pilate plainly and openly admits that he does not have a case against John's Jesus (18.38).

When the creases of Jesus' scarred wounds and his disciples' flesh press against each other in John's Gospel, the generated impression involves what Ahmed calls 'the cultural politics of emotion' (Ahmed, title) as people circulate and negotiate affective energy among one another in their intersubjective relating and relationship. This Johannine episode illustrates in a graphic way how the disciples' formation carries the pain and injustice suffered by John's Jesus. When John's text is pressed upon its readers, it may also impress upon them how colonized identity is connected with buried memory and history.

What if we read John's Jesus as still nursing open wounds on his hands and side?

Reading and Feeling the Open Wounds

In his piece about mourning and melancholia that has been so important for thinking about a politics of grief and mourning, Freud suggests that one of the 'traits' of a melancholic is 'an insistent communicativeness which finds satisfaction in self-exposure' (Freud vol. 14, p. 247). I have commented elsewhere that John's Jesus seems to suffer from 'a form of logorrhea' (Liew 2009, p. 260): not only does he talk a lot, but he also talks a lot about himself.

Freud writes, 'The melancholic are not ashamed and do not hide themselves, since everything derogatory they say about themselves is at bottom said about somebody else' (Freud vol. 14, p. 248). Homi K. Bhabha refers to this point by Freud, connects it with the work of Frantz Fanon, and writes:

> This inversion of meaning and address in the melancholic discourse – when it 'incorporates' the loss or lack in its own body, displaying its own weeping wounds – is also an act of 'disincorporating' the authority of the Master. Fanon ... says something similar when he suggests that the native wears his psychic wounds on the surface of his skin like an open sore – an eyesore to the colonizer. (Bhabha 1991, p. 102)

For John's Jesus, his injury is not just comparable but actually related to a physical laceration of the body. Since John's Gospel is clear that the Romans were the only ones with the power to kill (18.31) and that the wounds of John's Jesus came from the hands of the colonizing Romans, reading his wounds through Bhabha's reading of colonial wounds through Freud and Fanon can also add nuance to one reading of John's Gospel.

John's Jesus in this reading is returning as a colonized victim who bears and bares his wounds to not only protest against the colonizers but also transfer the feeling *of* the cross to his disciples, including Thomas (20.20, 27). He performs a 'show-and-tell' that is similar to Emmett Till's mother deciding to 'pass on' the 'cultural haunting' of being African American (Holloway 2003, p. 136) by having an open casket during her son's funeral, 'so that the world could see what they had done to [her] child' (cited in Holloway, p. 25; cf. Holloway, p. 130). This makes sense especially because his disciples, with the exception of the beloved disciple, may not have been present to witness the death of John's Jesus. Besides the fact that none of them is said to be present during the crucifixion scenes in John (19.25), John's Jesus has also announced that his disciples will scatter and abandon him during his 'hour' (16.32). Furthermore, the greeting of 'peace' that John's Jesus gives his disciples in both of his back-to-back appearances through closed doors (20.19, 26) should remind a careful reader of the disciples' desertion and hence absence from the foot of the cross. This is so because John's Jesus has assured them immediately after his announcement about their scattering at his 'hour' that they should still have 'peace', since his 'Father' will keep him company (16.32–33).

More than just authenticating his identity as their crucified, dead and now risen *and living* Lord, John's Jesus makes a point of flaunting his wounds to his disciples so they will, like the beloved disciple, feel him, feel with him, and identify with him. John's narrative is not shy about a 'transcorporeal and transformative' relation between its Jesus and Jesus' disciples (6.53–57; cf. Buell, pp. 71, 79–80). John is also clear that this relation involves affective transfer. We see this when John's Jesus is so moved by Mary's loss of her brother, Lazarus, that John's Jesus ends up weeping (11.28–36), even or especially when he knows that he can awaken Lazarus from his death (11.1–6, 11–15). Similarly, by showing his wounds to his disciples and to Thomas, John's Jesus ensures that they will feel, know, and remember that colonial loss and trauma are both personal and collective. In this reading, the open wounds of John's Jesus become infectious. By showing and opening up his crucified body to his disciples, John's Jesus opens them to see and feel the deep traumas of the colonial world under Rome. As Ahmed suggests, what we feel and what we do 'is shaped by the contact we have with others' (Ahmed, p. 4).

Referring to Freud's work on melancholia, loss and grief, Butler wonders about gender performance being done and 'understood as "acting out"' (Butler 1997, p. 145). The Buddhist monks and nuns in Vietnam who performed public suicides by fire (or by disembowelment) in Vietnam in the 1960s were clearly, to use Butler's words, 'acting out' (cf. Chiu 2009). Not only were these acts done in public, but a tip was also sent to an American news correspondent the night before the first of such suicides to ensure that the act would be captured and, then, circulated worldwide (Yang, pp. 1, 5–7). They were 'clearly theater[s] staged by the Buddhist monks to achieve a certain political end' (Browne 2003, p. 101). While media in the United States tended to be divided in framing these incidents as protests against either President Ngo Dinh Diem's pro-Catholic and anti-Buddhist policies or communist sympathizers working with the Viet Cong to destabilize the Republic of Vietnam (cf. Skow and Dionisopoulos 1997), one must remember that the Diem regime of South Vietnam was initially backed by the United States of America and its Catholic President (John F. Kennedy) to contain and combat the spread of communism. In other words, the so-called Buddhist crisis of 1963 in Vietnam can be read as a protest against *both* domestic oppression *and* colonial domination (whether through religious or state intervention, or both). Most importantly, as Freud and Bhabha have pointed out, these acts of self-immolation were actually 'performing a visual embodiment of violence done by an "other"' (Yang, p. 2). These public performances of suicide, these open and haunting displays of wounds and deaths should, therefore, be read as a staging of national and colonial grief; it is a staging that repeatedly demands 'something to be done' (Gordon 1997, pp. 139, 168, 183, 194, 202; 2011, pp. 1–3).

One suicide led to another and another, as more and more Buddhist clergy and ordinary citizens in Vietnam were exposed and awakened to their colonial pain and grief by what Bhabha calls an open sore and eyesore. After Thich Quang Duc killed himself by self-immolation on 11 June 1963, four more monks and a nun set themselves ablaze before a military coup. These suicides helped bring about the end of Diem's regime on 1 November 1963. Even after that, monks continued to commit suicide by self-immolation to protest the increasing presence of the United States in Vietnam under President Lyndon Johnson. In fact, a citizen of the United States, Norman Morrison, also burned himself to death in front of the Pentagon on 2 November 1965 to protest the Vietnam War, or what Vietnamese call 'the American War' (cf. Patler 2015). While Morrison's self-immolation is arguably best known, his was not the only one that took place in the United States during this war (King 2000, p. 128).

Without denying the important difference that these people took their own lives and were not directly killed by others like John's Jesus was, what I consider to be similar between them is the performative side of their respective stories, along with the affective transfers that result. By showing his pierced hands and side not once but twice (20.20, 27), John's Jesus literally engages in what Richard Schechner calls 'twice-behaved behavior' (Schechner 1985, pp. 35–6, 150 n.1) or performance. Since John's Jesus shows but makes no *overt* comment or gives any *specific* remarks about his wounds, his show(ing) may also be read in light of Benjamin's 'choreographic pantomime' (1998) that registers loss by 'bringing bodies to the foreground' (Butler 2003, p. 470).

The resurrected body of John's Jesus carries and exhibits the colonial violence covered up by *Pax Romana* with its open wounds. The fact that the resurrected body of John's Jesus continues to bear open wounds might also explain why Mary Magdalene and the other disciples have difficulty recognizing the resurrected Jesus (20.11–16; 21.4–8). People who have gone through trauma do change; they literally look different because their bodies now carry the wounds they suffered, especially if the trauma involves something like crucifixion. With John's Jesus volunteering to show his open wounds in these two post-resurrection appearances, the Fourth Gospel, like the picture of a burning Duc (cf. Yang, pp. 2–3), provides its readers with a frozen-in-time image of Rome's imperial terror and brutality by perpetuating the trauma experienced by John's Jesus as a literary spectacle. Thomas, as a result, personally and emphatically identifies (with) John's Jesus as '*my* Lord and *my* God' (20.28, NRSV). These words of identification may take on additional meaning if they are read as an allusion to Psalm 35.23, given that psalm's imprecatory plea for God to defeat and destroy the enemies of God's own people (cf. de-Claissé-Walford 2011).

John Ashton, despite his lack of interest in the body and the physical injuries of John's Jesus in this scene, writes, 'If John invented this story, as there is every reason to believe, it was not surely, to stimulate his readers to reflect upon the tangibility of risen bodies, but to *impress* upon them the need for faith' (Ashton 1991, p. 514; cited in Moss 2019, p. 135 n.5, emphasis mine). This is indeed an impressive scene, in Ahmed's sense of the word, because of Thomas's rather astonishing and aggressive demand, especially if one understands the text as referring to open wounds rather than scars: he wants to put his finger *in* the nail holes on the hands of John's Jesus, and then his (whole?) hand *into* the spear wound on the torso of John's Jesus (20.25). As if the Gospel writer is afraid that readers may miss the picture, the narrative has John's Jesus basically repeating Thomas's words to Thomas when he appears to him (20.27). These words should bring up for a reader 'visceral sensations of revulsion and disgust' (Most 2005, p. 49). Moreover, these repeated words may remind readers that the nail and spear wounds borne by John's Jesus are caused by human hands (19.13–18, 23, 31–34). The passage functions, then, to align the feelings of the Gospel's readers *with* the pain of John's Jesus and *against* the aggression or bloodlust of his assailants. Since these with-and-against feelings are associated with Roman crucifixion in particular and imperial violence in general, they should provoke also feelings of dread and perhaps even terror about the future.

Reading and Feeling Deaths

The impression one gets from this scene may become even heavier in light of what follows. What we basically have are two more scenes through which John's Gospel impresses upon its characters and its readers the reality of death and dread. First, John's Jesus tells Peter that he will lose his freedom, with the narrator quickly clarifying that John's Jesus is actually referring to his death (21.18–19). Even the call for Peter to 'feed' or 'tend' the sheep of John's Jesus (21.15–17) is suspect, given not only the fate of Peter but also the death of John's Jesus as the Passover lamb. Second, through a conversation between Peter and John's Jesus, the narrator seems to explain for its readers the beloved disciple's death or help them make sense of it (21.20–23). The resurrection of John's Jesus only ends with frustrated expectations and more references to existing scars/wounds and future death; there is no real resolution or reconciliation to wrap up John's Gospel. If anything, the ghostly appearances of John's Jesus, the 'foretelling' of Peter's demise, and the mention of the beloved disciple's death may cause its readers to think about 'what comes after loss for the survivors' (Kim 2019, p. 55).

By showing his scars and/or wounds to his disciples and then to Thomas, John's Jesus is 'acting out' a performance that 'makes visible (for an instant, live, now) that which is always already there: the ghosts, the tropes, the scenarios that structure over individual and collective life' (Taylor 2003, p. 143). These messages of death, one after another after another, bring up the larger picture of – and Williams's 'structure of feeling' (1975) for – people who died under Roman colonialization who needed to be remembered, recognized and mourned. John's Gospel testifies in a sense to its community's vulnerability to colonial carnage, as Caiaphas' comment (11.45–53) clearly shows. These closing episodes in John 20—21 give us a Jesus who, despite his repeated references to his imminent departure (7.33–34; 8.21; 12.35–36; 13.33, 36; 14.1–4, 12, 18–19, 25–28; 16.5–11, 16–19, 28; 20.17), actually refuses to disappear and, without any definite closure, continues to spectralize. This is so because, by giving us two conclusions (20.30–31; 21.24–25), John seems to struggle with his 'minor feelings [that] are ongoing' (Hong 2020, p. 57), so he has difficulties in closing his Gospel. In addition, these chapters may serve as a way for John's readers to *also* feel the suffering and death of John's Jesus through what I am calling affective transfer. '[T]hose who have not seen and yet have come to believe' (20.29, NRSV) may hence be referring to those whom John's Jesus calls 'those who will believe in me through [his disciples'] word' (17.20, NRSV).

With the term 'postmemory', Marianne Hirsch proposes that later populations who have not experienced the Holocaust directly may nevertheless, through a process of *retrospective witnessing by adoption* – have the capacity of 'adopting the traumatic experiences – and thus also the memories of others – as experiences one might oneself have had' (Hirsch 2001, p. 10, emphasis original; cf. Hirsch 2012). These chapters, if I may borrow Jermaine Singleton's words (2015, p. 51) about a play by August Wilson, show how grief 'is transferred ... as a result of and in resistance to an enduring struggle with ... oppression'. John's Gospel ends, then, by passing on an ongoing grief to its readers. Readers are now supposed to carry on the memory of Jesus by, in John's language, 'testifying' (1.7–9, 15, 32–34; 3.11, 26, 32–33; 4.39; 5.31–33, 36–37, 39; 8.14, 17–18; 10.25; 12.17; 15.26–27; 18.37; 19.35; 21.24).

According to Ahmed, 'How we feel about others is what aligns us with a collective, which paradoxically "takes shape" only as an effect of such alignments. It is through how others impress upon us that the skin of the collective begins to take shape' (Ahmed, p. 54). Mourning for injuries, death and loss of John's Jesus – and the transfers of affect and emotion that it entails – can result in a political alliance not only among many

Johannine characters but also among readers of the Gospel. The appearances of John's resurrected but unhealed Jesus, with his scarred and/or wounded body, haunt us and give us hope at the same time.

Conclusion

The importance of grief and mourning in John's Gospel can be seen in the scenes with Mary Magdalene at the empty tomb (20.1–18). Although Peter and the beloved disciple accompanied Mary to the empty tomb and entered the empty tomb without her, they were visited by neither the angels nor John's Jesus. Glenn W. Most makes an important suggestion that it's Mary weeping that leads to the appearance of both the angels in the empty tomb and the resurrected Jesus himself (Most 2005, pp. 35–6). Note how John's narrative refers to Mary's weeping repeatedly (20.11, twice), and has both the angels and John's Jesus asking her about the reason for her weeping (20.13, 15). Mary's response to them further shows that her tears have to do with the missing body of John's Jesus (20.13, 15). In other words, she is weeping for her inability to mourn her loss of John's Jesus properly without his body, the absence of which 'brings home to her in an especially distressing way her irrevocable loss' (Most, p. 36). Contrary to Peter and the beloved disciples (20.10), Mary Magdalene simply refuses to leave the tomb even though it is empty and she is there alone again, just as she went to the tomb all by herself in the early morning supposedly to mourn the passing of John's Jesus. It is *her grief* – in fact, her desire and her determination to mourn properly – that brings about not only the first actual appearance of angels in John's Gospel but also the resurrected appearance of John's Jesus after his death and burial.

The resurrection of John's Jesus in my reading has more to do with transmitting colonial trauma and colonial grief; it is a form of protest against colonialism through a demonstration of scars/wounds than a process, a promise, or a possibility of healing. With John's realized eschatology, there is little indication that John's resurrected Jesus is coming back to set things right. There is no teleological resolution or a crowning closure. Instead of returning to an 'old normal' or settling with a 'new normal', I have suggested reading John as a piece of colonized and minoritized writing *for* and *as* a politics of grief and mourning. Such a reading makes a political claim and carries the potential, through the transmission of affect, to foster the building of a group identity. It is not a politics of passivity and resignation, but one that may motivate readers by confronting them persistently with previously unacknowledged or under-recognized losses suffered by colonized and minoritized subjects. It attests to how our narratives – biblical, national or global – are broken and full of brokenness.

A poetics and politics of grief and mourning, by keeping alive those 'what-could-have-been' alternatives (Gordon 1997; 2011) and 'what now' questions (Yang 2011, p. 16), would be my unhopeful hope in passing on the haunting pasts. Perhaps this ancient Gospel called John can still help us remember and reimagine with its mourning of a ghostly Jesus figure in these times when most of the world seems to be taken over by not only a pandemic but also the pandemonium of the United States Empire.

References

Sara Ahmed, 2014, *The Cultural Politics of Emotion*, 2nd edn, Edinburgh: Edinburgh University Press.

John Ashton, 1991, *Understanding the Fourth Gospel*, Oxford: Oxford University Press.

Walter Benjamin, 1998 (German original, 1963), *The Origin of German Tragic Drama*, trans. John Osborne, New York: Verso.

Homi K. Bhabha, 1991, 'A Question of Survival: Nations and Psychic States', in James Donald (ed.), *Psychoanalysis and Cultural Theory: Thresholds*, pp. 89–103, New York: St. Martin's.

Malcolm W. Browne, 2003, 'Vietnam, Persian Gulf', in Michelle Ferrari (ed.) with commentary by James Tobin, *Reporting America at War: An Oral History*, pp. 91–110, New York: Hyperion.

Denise Kimber Buell, 2014, 'The Microbes and Pneuma That Therefore I Am', in Stephen D. Moore (ed.), *Divinanimality: Animal Theory,*

Creaturely Theology, pp. 63–87, New York: Fordham University Press.

Judith Butler, 1997, *The Psychic Life of Power: Theories in Subjection*, Stanford: Stanford University Press.

———, 2003, 'Afterword: After Loss, What Then?', in David L. Eng and David Kazanjian (eds), *Loss: The Politics of Mourning*, pp. 467–73, Berkeley: University of California Press.

———, 2004, *Precarious Life: The Powers of Mourning and Violence*, New York: Verso.

Anne Anlin Cheng, 2001, *The Melancholy of Race: Psychoanalysis, Assimilation, and Hidden Grief*, New York: Oxford University Press.

Lily V. Chiu, 2009, '"An Open Wound on a Smooth Skin": (Post)Colonialism and the Melancholic Performance of Trauma in the Works of Linda Lê', *Intersections: Gender and Sexuality in Asia and the Pacific* 21 (accessed 23.3.21: http://intersections.anu.edu.au/issue21/chiu.htm#t29).

Ann Cvetkovich, 2003, *An Archive of Feelings: Trauma, Sexuality, and Lesbian Public Cultures*, Durham: Duke University Press.

Nancy L. de-Claissé-Walford, 2011, 'The Theology of the Imprecatory Psalms', in Rolf A. Jacobson (ed.), *Soundings in the Theology of the Psalms: Perspectives and Methods in Contemporary Scholarship*, pp. 77–92, Minneapolis: Fortress.

W. E. Burghardt DuBois, 1907, *The Souls of Black Folk: Essays and Sketches*, Chicago: A. C. McClurg & Co.

David L. Eng and Shinhee Han, 2000, 'A Dialogue on Racial Melancholia', *Psychoanalytic Dialogues* 10.4: 667–700.

David L. Eng and David Kazanjian, 2003a, 'Introduction: Mourning Remains', in David L. Eng and David Kazanjian (eds), *Loss: The Politics of Mourning*, pp. 1–25, Berkeley: University of California Press.

——— (eds), 2003b, *Loss: The Politics of Mourning*, Berkeley: University of California Press.

Yên Lê Espiritu, 2014, *Body Counts: The Vietnam War and Militarized Refugees*, Berkeley: University of California Press.

Sigmund Freud, 1953–74, *Standard Edition of the Complete Psychological Works of Sigmund Freud*, 24 vols, ed. and trans. James Strachey, London: Hogarth.

Avery F. Gordon, 1997, *Ghostly Matters: Haunting and Sociological Imagination*, Minneapolis: University of Minnesota Press.

———, 2011, 'Some Thoughts on Haunting and Futurity', *borderlands* 10.2: 1–21.

Donna J. Haraway, 2016, *Staying with the Trouble: Making Kin in the Chthulucene*, Durham: Duke University Press.

Marianne Hirsch, 2001, 'Surviving Images: Holocaust Photographs and the Work of Postmemory', *The Yale Journal of Criticism* 14.1: 5–37.

————, 2012, *The Generation of Postmemory: Writing and Visual Culture after the Holocaust*, New York: Columbia University Press.

Karla F. C. Holloway, 2003, *Passed On: African American Mourning Stories*, Durham: Duke University Press.

Cathy Park Hong, 2020, *Minor Feelings: An Asian American Reckoning*, New York: Oneworld.

Sara Clarke Kaplan, 2007, 'Souls at the Crossroads, Africans on the Water: The Politics of Diasporic Melancholia', *Callaloo* 30.2: 511–26.

Jinah Kim, 2019, *Postcolonial Grief: The Afterlives of the Pacific Wars in America*, Durham: Duke University Press.

Sallie B. King, 2000, 'They Who Burned Themselves for Peace: Quaker and Buddhist Self-Immolators during the Vietnam War', *Buddhist-Christian Studies* 20: 127–50.

Tat-siong Benny Liew, 2009, 'Queering Closets and Perverting Desires: Cross-Examining John's Engendering and Transgendering Word across Different Worlds', in Randall C. Bailey, Tat-siong Benny Liew and Fernando F. Segovia (eds), *They Were All Together in One Place? Toward Minority Biblical Criticism*, pp. 251–88, Atlanta: Society of Biblical Literature.

————, 2016a, 'Haunting Silence: Trauma, Failed Orality, and Mark's Messianic Secret', in Tat-Siong Benny Liew and Erin Runions (eds), *Psychoanalytic Mediations between Marxist and Postcolonial Readings of the Bible*, pp. 99–127, Atlanta: SBL Press.

————, 2016b, 'The Gospel of Bare Life: Reading Death, Dream, and Desire through John's Jesus', in Tat-Siong Benny Liew and Erin Runions (eds), *Psychoanalytic Mediations between Marxist and Postcolonial Readings of the Bible*, pp. 129–70, Atlanta: SBL Press.

Candida Moss, 2019, *Divine Bodies: Resurrecting Perfection in the New Testament and Early Christianity*, New Haven: Yale University Press.

Glenn W. Most, 2005, *Doubting Thomas*, Cambridge, MA: Harvard University Press.

Nicholas Patler, 2015, 'Norman's Triumph: The Transcendent Language of Self-Immolation', *Quaker History* 104.2: 18–39.

Edward W. Said, 1993, *Culture and Imperialism*, New York: Alfred A. Knopf.

Richard Schechner, 1985, *Between Theater and Anthropology*, Philadelphia: University of Pennsylvania Press.

Jermaine Singleton, 2015, *Cultural Melancholy: Readings of Race, Impossible Mourning, and African American Ritual*, Urbana: University of Illinois Press.

Lisa M. Skow and George N. Dionisopoulos, 1997, 'A Struggle to Contextualize Photographic Images: American Print Media and the "Burning Monk"', *Communication Quarterly* 45.4: 393–409.

Ann Laura Stoler (ed.), 2013, *Imperial Debris: On Ruins and Ruination*, Durham: Duke University Press.

———, 2016, *Duress: Imperial Durabilities in Our Time*, Durham: Duke University Press.

Diana Taylor, 2003, *The Archive and the Repertoire: Performing Cultural Memory in the Americas*, Durham: Duke University Press.

Raymond Williams, 1975 (original, 1961), *The Long Revolution*, Westport: Greenwood .

Joseph R. Winters, 2016, *Hope Draped in Black: Race, Melancholy, and the Agony of Progress*, Durham: Duke University Press.

Michelle Murray Yang, 2011, 'Still Burning: Self-Immolation as Photographic Protest', *Quarterly Journal of Speech* 97.1: 1–25.

25

'Today I Let My Rage Be Beautiful': Po/et(h)ical Responses to the New Normal

DOROTHEA ERBELE-KÜSTER

Po/et(h)ical responses are powerful ways to face the absurdity of what is called the 'New normal'. They bind the aesthetic and the political together, the personal and the social. This is captured by the wordplay Po/et(h)ics.

The lockdown due to the Covid pandemic gave birth to new artistic formats in the digital world. The pandemic likewise impacted our language by the flood of new terms[1] that determine the perception of the crisis. The way we talk about the crisis, even the virus itself, influences how we deal with it. This can be traced also in the recent protest movements around the globe which were accompanied by artistic expressions by individuals and communities. This is expressed in the fulminating opening line of Drew Drake's poem 'Today I let my rage be beautiful' which serves as the title of this chapter. The poem reflects the role that affective emotions have in the response to the ongoing discrimination and violence faced by the poet and the black community he is part of.

Next to this poetical response to the so-called new normal from an African American context rooted in the Black Lives Matter movement I shall discuss two more responses to the crisis in the spring and early summer 2020 stemming from the European context. Each artist felt urged to respond with his/her means – poetics. Special features of the poetical discourse such as wordplays, rhythm and sound have to be taken into account as they incite pleasure, disgust, compassion and other emotions. Likewise, they respond to emotions as perception of the world.

The affective mode of perceiving the world is considered crucial in the perception. One emotion and po/et(h)ical expression 'anger' strikes most: under the form of lament, the disguise of close description, accompanied by longing for change. The poems range from prophetic description to lament and protest. I shall sketch the link between lament and rage and the close description of the situation which stands at the beginning of every meaningful poet(h)ical expression. I read these poetical responses as a scholar trained in interpreting biblical texts and with a special interest in the poetical form and its specific communicative and rhetorical power in coping with the crisis.[2] At the backdrop lies the question of understanding the contribution that each of the diverse poetical forms can make. This gives us insights for the interpretation of biblical texts in times of crisis.

'The Invisible Giant': Danny Dziuk

Danny Dziuk, a composer and songwriter in Berlin,[3] stands in the German tradition of Liedermacher. The song 'Der strahlend blaue Himmel dieser Tage' / 'The bright blue sky of these days' describes his perceptions of the glocal situation in spring of 2020.[4] His voice is accompanied with an elegiac guitar sound. The song was composed during the lockdown in April under an extremely blue sky with almost no airplanes and enchanting spring blossoming.[5] Dziuk explains in an interview shortly after the release of the song that facing the pandemic, he felt helpless and speechless and not at all pushed to write a Corona-song. A phrase he overheard from Alexander Kluge, a German writer and filmmaker, commenting on the blue marvellous spring sky urged him spontaneously to compose.[6] The idyllic blue sky becomes the starting point for Dziuk's uneasiness and sceptical questioning as expressed in the refrain (my translation):

> The bright blue sky of these days
> what is it that is wrong with it?
> While an invisible giant takes the world's breath away and
> ends its party.[7]

The blue sky serves as a mirror for the situation in Germany and other parts of the world that were abruptly slowed down by the lockdown. Danny Dzuik deliberately avoids an appraisal of the bright blue sky and the less polluted air in order not to trap in the pitfall of appraising the side-effects of the virus, especially the reduction of the traffic on the streets and in the air due to the lockdown. But the bright blue sky lays bare what goes wrong:

> Some can easily afford the lockdown
> others hardly get out of their debts
> and those who sweat away
> do not pay their rent simply with applause

Longstanding burning issues become visible through the lens of the lockdown. He calls them 'the wrong decisions of the last years' – the health care system, the industrial production of food, economic crisis, social inequality, migration, and last but not least the climate trauma as the lockdown seems to function confusingly as a pause button.

> the gold price is rising to unprecedented heights ...
> rigorous police checks in New Delhi to military operations
> in South Africa ... –
> thousands are quickly brought from Eastern Europe to
> harvest asparagus [in Germany], and for parentless
> children from Camp Moria the capacities are exhausted
> The bright blue sky these days
> what is it what is wrong with it
> while a giant shows us ... the moral bankruptcy.

The composer does not give room to exuberant emotions. He is cautious to draw conclusions. He questions himself and the listeners. He takes his time for the close description of his view out of his window. We listen for six minutes to his measured, smooth, melancholic voice and the electric guitar sound.

The song does not give homage to Covid-19 as it consequently avoids all the c-terms. It does not want to be reduced to a Corona-song. The small virus is surnamed with a compelling metaphor: the giant, the invisible one. The metaphor personalizes the Sars-CoV-2 virus by upholding its inconceivability. Half a year later, as I write this essay in late autumn 2020, this has become true moreover. The invisible giant, only perceivable through a microscope, has overwhelmed us and pervades daily and public life in Germany, even more than in spring. In the media we see two pervasive strategies to deal with 'the invisible giant' – one responds to the invisibility by trying to depict the virus with symbols. 'Visualizing the virus has been both panacea and political tool' (Chatterjee 2020).[8] The designs serve as a kind of taming the gigantic power of the small virus which is invisibly transmitted mostly through aerosol. The other strategy is to cover the virus up with statistics. The loss and the daily struggle of the medical personnel is hidden in death tolls. Although the virus stays invisible it makes visible the crisis and our 'moral bankruptcy'.[9] Reflecting under the blue clear sky Dziuk discloses with a prophetic voice what the invisible giant unveils while itself staying invisible. According to him the invisible giant does not have a message; he just takes our breath away.

'In search for a place for those who died alone': Ulrike Bail

Our difficulties with breathing resonate likewise in the poem by Ulrike Bail who has lived in Luxembourg since 2005.[10] Her latest anthology of poems came out in autumn 2020, whereas she started her career as a protestant theologian in Germany and wrote her dissertation on the Psalms (1998). The poem gropes for a place for loss and unnamed grief. She is in search for verbalization of the unspeakable. Just one word makes reference to the actual pandemic – Bergamo – which has become a grieving symbol of the pandemic in Europe. Her longing for the other is expressed through speechless lament (my translation):

> the winter in your mouth through the empty city
> in your hand the locked-out breath
> prayers like disposable towels or undergrowth
> in search for a place for those who died alone
> perhaps a pear bosk a high beech
> I would call this stop Bergamo
> Bergamo arrêt supprimé[11]

In the dense poem, the sentences are sometimes broken. They seem unfinished as either verb or subject is missing. At the end of the poem a subject emerges through the voice of the first person singular. The concrete is used in a metaphorical way in the poem. Are we riding through a town in a bus or are we walking through the bush? With a mix of pictures, we likewise have to deal when it comes to prayer. Two uncommon metaphors are used: one stems from the hygienic, the other from fauna. Prayers are called disposable towels for single use only. Disposable towels are purifying, they wipe away. Likewise, prayers are as bushes in a deserted landscape where everything dies back. The I-voice would call the place for prayers: 'Bergamo'. The city in a highly populated region close to Milan in North Italy, an industrial centre, has suffered from a high mortality rate due to Covid-19. Bergamo is the name given to the city on a hill at the base of the Alps by the Celts: Bergheim – Bergamo – Home on the hill. It resonates with the description of the place by Ulrike Bail as 'Bergbirnenhain' making use of an alliteration: a pear bosk (pear bush) on a hill. However, in the poem, this place for rest and lament is a stop which is not served. The place for lament is suppressed. She chooses a French expression of her new multilingual home, Luxembourg. The requested stop is not responded to. No stop. No rest. Our journey does not halt. No time and place for lament.

In Upper-Italy and around the globe in spring 2020, almost only the gravediggers stood at the graves to mourn. The cry of lament could only be heard in the ringing of death bells. Grief-struck families mourned their coronavirus dead online as they were denied a funeral. They shared photos, memories and sorrow that their loved ones had died alone. This has turned into a protest movement trying to persecute the ones they hold responsible for the deaths: *Noi Denunceremo* (We will denounce) *Verità e giustizia per le vittime di Covid-19* (Truth and justice for the victims of Covid-19). Against this background the poem is a plea for grievability.

'Searching my rage': Drew Drake

The final poem to discuss was composed by Drew Drake, a New York based teaching artist. It emerged from protest movements including Black Lives Matter and People's strike fighting against Covid-Capitalism.[12] The difficulty of breathing freely is addressed: 'Realized what a privilege it is to breathe this GOOD air,/ When many of my family members have died.' These lines echo 'I can't breathe', the outcry repeated by George Floyd as his breath and life was extinguished by a police officer kneeling on his neck on 25 May 2020. The rap poetry[13] responding to racism and state sanctioned violence explicitly addresses emotions, in particular rage, which seems to embrace all the poet's longings, his pain and his pleasure. It is an emotion sensed by many of his community and around the globe.

Emotions in general as embodied expressions play an important role in moral decision making as they seek to respond to situations that cannot be dealt with in a routine way (Mesquita and Frijda 2011). The poem tries to understand the emotional state facing the new normal and its brutal injustice. It expresses that emotions are not mere inner feelings but that they are interpersonal as they respond to situations. It gives room to the multiple shapes of anger declaring: 'Today / I let my rage be beautiful.'

Rage, as a usually disregarded emotion in ancient philosophy as in modern psychology, is acclaimed as beautiful and no longer suppressed or dressed up. According to ancient philosophers anger only makes sense for those in power. Yet, it rises exactly due to the lack of reciprocity. Audre Lorde, the black feminist activist, has phrased this in her 1981 keynote speech as follows: 'Women responding to racism means women responding to anger; Anger of exclusion, of unquestioned privilege, of racial distortions, of silence, ill-use, stereotyping, defensiveness, misnaming, betrayal, and co-optation' (Lorde, p. 278). Anger and indignation are likewise prevalent in other current protest movements around the globe in their striving for justice.[14] The philosopher Amia Srinivasan argues in a recent article for the aptness and productiveness of anger (2018). Anger may serve as a means by which one can come to better see one's oppression. Getting angry is a way of 'affectively registering or appreciating the injustice of the world' (Srinivasan, p. 132). Along with other feminists she calls this 'affective injustice' bringing to the core that getting angry is more than just knowing the injustice. The affective mode of perceiving the world is thereby considered crucial. In his poem, Drake alludes to other songs stating that his anger may sound like Knuck If You Buck,[15] Amazing Grace,[16] and Lift Every Voice and Sing.[17]

Drew Drake hints to a huge reservoir of songs ranging from Psalm prayers, Gospels to Crunk music, that provide guidance in the struggle for survival.[18] The music calls the tune by making audible an orchestra of voices and by doing so it gives the people their voice back. Performing the song situates the current moment in the history of the struggle for freedom over several centuries. The songs serve as a medium for survival. The question at the end – 'And Don't it sound like this rage is Total Praise?' – suggests that all the polychord voices are finally a hymn jubilating justice and beauty. It makes me think of the theology of the Psalter. The Psalter ends on universal praise without annihilating lament and protest against injustice: 'Everything breathing shall praise Jah! Hallelujah' (Ps. 150). Reading Psalms is breathing as we all breathe and are in need of praying, humming, singing and breathing in the midst of crisis (Erbele-Küster 2020).[19]

I would like to identify two more theological allusions to

how rage is characterized: first, 'Today – I let my rage be Holy.' Anger by humans is seen negatively in the Bible. 'Good' anger is reserved for God. An exception might be Moses expressing anger after the speech of Pharaoh who refuses to let the Israelites leave bondage (Ex. 11.8).[20] A second allusion to a biblical text that Drew Drake makes is to the creation account when he proclaims: 'I let it be. And it was Beautiful.' This echoes God's final statement at the end of his creative acts in Genesis 1: 'And God saw: It is good.' The Hebrew uses a word underlying the aptness of something and which encompasses goodness in the sense of beauty (cf. Erbele-Küster 2019). Against the general negative perception of rage as destructive it becomes a creative act as the artist-and-activist become the creator stating: 'I let it be. And it was Beautiful.'

Po/et(h)ics Beyond the New Normal

All poets face the so-called New Normal with their master tools – poetics. They are united in their indignation about the situation resulting in a compassionate response. The poems can inspire us in our endeavours as theologians in at least a two-fold way. Reading them as late-modern Psalms hints at three distinctive and yet related po/et(h)ical ways of responding to the crisis. This may also help us discern the diverse biblical po/et(h)ical literary forms.

Danny Dziuk acts like a prophet exposing injustice, moral ambiguities and the absurdities we faced during the lockdown. Neither heaven nor a utopian place but the blue clear sky serves as a striking horizon. From an environmental perspective one could welcome the virus and its consequences – Dziuk does not draw such a conclusion and he avoids the danger of moralistic misuse. And still the invisible giant lays bare the moral bankruptcy including our endless pollution of the sky and the waters. The close description by Dziuk ascertains that we do not lose out of sight the wide range of issues we have to cope with which tend not to be covered by the media as we focus solely on Covid-19. 'The strange blue sky of these days' reflects the malady the earth suffers beyond the invisible virus. In Albert Camus' *La Peste* Doctor Rieux is asked if he thinks that the plague has positive side effects as opening the eyes. He shakes his head and points to the disaster, the sufferance due to the malady.

The verbalization of the loss stands at the backdrop of the poem by Ulrike Bail. In order to commemorate the dead, the I-voice in the poem searches desperately for a place for lament and rest. In this the poem reflects on how prayer may look like in order to make a contribution in dealing with the loss. It is a subtle search to express the grievability of every human being, reminding us of a central biblical theological aspect: to voice the vulnerability of our human existence in communal prayer.

Drew Drake powerfully mixes up biblical and postmodern Hip-Hop language, turning the allusions to biblical texts and common norms upside down. Rage becomes a creative act. Cheryl Kirk-Duggan (2017, p. 38) states in a similar way: 'anger or rage can be creative and is a choice; depicts grief and loss as an in-between place of anguish and pain that often follows anger, and precedes healing and completion. Such fury and sorrow also celebrate the journey through rage and grief toward hope, transformation, and an ultimate covenantal relationship with God.' This underlies the relation between lament and anger.

Searching my rage conceives a close relation between anger, passion and compassion. The moral companions of anger, indignation and compassion are rooted in our corporal existence, our relation to the other; they show us the abyss of social distancing. The erring of the I-voice in 'Bergamo arrêt supprimé' who has difficulty breathing is full of compassion. Poet(h)ics ranging from close self-critical description, affective passion and breathless lament express our vulnerable existence and uphold us by reminding us of the beauty of it.

Notes

I dedicate this contribution to my friends – named and unnamed.

1 In the German-speaking context already more than 1,000 neologisms have been catalogued by the Institut für Deutsche Sprache in Mannheim – IDS: Neologismen in der Coronapandemie (accessed 22.12.20: www1.ids-mannheim.de/neologismen-in-der-coronapande mie/).

2 I have started to classify the poetical forms within so-called biblical crisis literature. I distinguish mainly three models ranging from stumbling voices mostly of lament within crisis, and critical prophetic account to (moral) reimagination beyond crisis.

3 Born in 1956 in Duisburg, Danny Dziuk lives in Berlin, Germany (see http://dziuks-kueche.de/).

4 Available at www.youtube.com/watch? v=2-5sc8OzfDk. The song was broadcast by the free radio channel in Berlin and Radio Berlin Brandenburg. See an interview with him after the release of the song (accessed 20.12.20: www.radioeins.de/programm/sendungen/mofr1921/ interviews/danny-dziuk.html).

5 Eva Horn links to each season a specific imaginative mode of the (climate) catastrophe (see chapter 2, Catastrophe Without Event: Imagining Climate Disaster). In Germany, for instance, we confronted in late autumn and winter 2020 a second lockdown – the psychological and social perception of the pandemic became aggravated, whereas in the first lockdown the sensation of spring and blossoming has been counterintuitive to the lockdown and the threat of being infected by the virus.

6 See www.monopol-magazin.de/interview-alexander-kluge-corona.

7 Used by permission of Danny Dziuk.

8 Helmut Renders (2021 forthcoming) compares the visualization of the virus with vanitas depictions and concludes that they serve as a kind of 'Andachtsbild' mirror for meditating on the virus and its possible deathly consequence. Through visualization the virus and the fear of it is transformed into an artefact.

9 Bernard Levy argues vehemently that the virus itself does not interpret nor has a message. In a sense his verve is a polemic against moral statements in general and tends itself to play down our failures.

10 See www.ulrike-bail.de/Ulrike_Bail/Willkommen.html.

11 The poem is part of a new anthology 'statt einer ankunft' (forthcoming in 2021 with Conte Verlag) which has just been awarded first place by the national committee in Luxembourg (accessed 23.3.21: https://luxembourg.public.lu/fr/societe-et-culture/creation-artistique/ concours-litteraire-national.html). Used by permission of Ulrike Bail.

12 See https://lifejacket.medium.com/meet-teaching-artist-drew-drake-9d21e53a7d73 (accessed 20.12.20).

13 Available at https://youtube/LwgD7Gg64CM (accessed 29.12.20). The poem was part of an online event on 8 June 2020 organized by Interfaith Youth Core together with Hebrew College 'Psalm Season'. The organizers state on the homepage about the project: 'We are living in a time of widespread illness, ongoing racism, and deep fear and division. People throughout the world are crying out, raising their voices in protest and lament, seeking hope and solace. Our voices echo those over the millennia who have cried out in every generation, turning to their spiritual traditions for guidance and inspiration' (accessed 20.12.20: https://ifyc.org/article/intro-project-time-upheaval). For an interreligious reading of the Psalms see Frevel 2020.

14 In Italy f.ex.: www.noidenunceremo.it; in Spain since 2011: Indignados movement.

15 This is the title of a legendary song by Hip Hop band Crime from 2004 expressing the wish to fight if you are offended.

16 This alludes to President Barack Obama singing the gospel in the funeral ceremony for the Charleston church attacks in 2015.

17 The black African American anthem.

18 Cheryl Kirk-Duggan (2020) likewise underlies the role aesthetics and music plays in her strive for justice and hence her womanist theology.

19 Rabbi Aaron Weininger searching for a prayer while attending the funeral service for George Floyd understands the prayer 'I can't breathe' on the backdrop of this final tune of the Psalter (accessed 23.3.21: www.jta.org/2020/06/05/opinion/i-am-a-rabbi-who-attended-george-floyds-memorial-service-what-i-heard-broke-me).

20 Baloian, p. 54: 'The text appears to justify his anger in light of the constant refusal of Pharaoh to listen, and he does not do more than angrily speak his mind.' McCaulley (2020) states that the Psalms are not silent about the rage of the oppressed.

References

Ulrike Bail, 1998, *Gegen das Schweigen klagen. Eine intertextuelle Studie zu den Klagepsalmen Ps 6 und Ps 55 und der Erzählung von der Vergewaltigung Tamars*, Gütersloh: Gütersloher Verlagshaus.

Bruce Edward Baloian, 1992, *Anger in the Old Testament*, New York/Bern/Paris: Peter Lang.

Sria Chatterjee, 2020, 'Making the invisible visible: How we depict Covid-19', LSE, 10 June (accessed 20.12.20: blogs.lse.ac.uk/impactofsocialsciences/2020/07/10/making-the-invisible-visible-how-we-depict-covid-19/).

Drew Drake, 2020, 'Searching My Rage' (accessed 31.3.21: www.you tube.com/watch?v=LwgD7Gg64CM)

Dorothea Erbele-Küster, 2019, 'Senses Lost in Paradise? On the Inter-relatedness of Sensory and Ethical Perceptions in Genesis 2–3 and Beyond', in Annette Schellenberg/Thomas Krüger (eds), *Sounding Sensory Profiles in Ancient Near East*, pp. 145–60, Atlanta: Society of Biblical Literature.

———, 2020, 'Reception Aesthetics of the Psalms. A Third Space for Intercultural and Interreligious Dialogue', in Christian Frevel (ed.), *Mit meinem Gott überspringe ich Mauern/By my God I leap over a wall*, Interreligiöse Horizonte in den Psalmen und Psalmenstudien, Herders Biblische Studien 96, pp. 415–32, Freiburg: Herder Verlag.

Christian Frevel (ed.), 2020, *Mit meinem Gott überspringe ich Mauern/By my God I leap over a wall*, Interreligiöse Horizonte in den Psalmen und Psalmenstudien, Herders Biblische Studien 96, Freiburg: Herder Verlag.

Eva Horn, 2018, *The Future as Catastrophe. Imagining Disaster in the Modern World*, New York: Columbia University Press.

Cheryl Kirk-Duggan, 2017, *Baptized Rage, Transformed Grief: I got through so you can*, Eugene, OR: Wipf & Stock.

———, 2020, 'Ruminating on the Color Purple. Womanist Engagement in the Time of 46-1 in the Oval, *Lectio difficilior* 2 (accessed 23.3.21: www.lectio.unibe.ch/20_2/cheryl_kirk_duggan.html).

Bernard Levy, 2020, *Ce virus qui rend fou*, Edition Grasset: Paris.

Andre Lorde, 1997 (reprint from 1981), 'The Uses of Anger', *Women's Studies Quarterly* 25: 278–85.

Esau McCaulley, 2020, 'What has the Bible to say about Black Anger', *New York Times*, 14 June (accessed 29.12.20: www.nytimes.com/2020/06/14/opinion/george-floyd-psalms-bible.html?action=click&-module=Opinion&pgtype=Homepage).

Batja Mesquita and Nico H. Frijda, 2011, 'An emotion perspective on emotion regulation', *Cognition and Emotion* 25: 782–84.

Helmut Renders, forthcoming, 'Culturas visuais religiosas em tempos de pandemia: uma leitura warburguiana das Danças dos mortos e das Vanitas', *PLURA. Journal of the Study of Religion, Juiz de Fora*.

Amia Srinivasan, 2018, 'The aptness of anger', *The Journal of Political Philosophy* 26: 123–44.

26

nonplussed

confusion is heard
in the contemplation of normal
the struggle to reconcile
a change from what was
the struggle to imagine
that the named normal of the past
which was permeated with injustice
could yield way to new life
and a new way of living

normal was a child of patriarchy
a spawn of injustices that manifested
as racism
as sexism
as colonialism
as enterprises
as imperialism
as greed
as death

making humans into machines
wounding men, women and children
creating death of mind, body and spirit
to yield wealth and gain
leaving scars and wounds
generational trauma
tyranny and brutality
an infectious oozing
spreading as pandemics

normal was death and more death
experienced in these pandemics
amplified by this Covid-19
which has peeled back layers
of miscarried justice
exposing the fissures
long named and unaccounted for
amidst the quest to keep alive a haunting past
sustaining the pandemonium of decaying empires

we weep
hearts broken with grief
tears for those who died alone
tears for those ignored
we rage
prophetic rage
a beautiful defiance
embraced as resistance to suppression
we reject the idea of returning to normal

daring discernment
calls forth
a reality that means
we pluck up
we pull down
we plant
we overthrow
we rebuild
rejecting notions of normal

why return to normal?
promoting privileges
where normal
multiplies the marginalized
where normal
ignores the least of these
where normal says there is
nothing to grieve
this destructive normal itself a pandemic

who among us is surprised?
at the rage in the streets
at the resistance worn globally
at the need to push back
at the burning bushes
the signs that all is not well
a sign that the Divine calls
calling out from a global lockdown
calling for an upheaval to normal

there is a new thing pushing up
a new thing springing out among us
new language reflecting God talk
holding on to the Divine among us
when we cannot mourn
what do we imagine?
what do we hear of God?
in the mourning
in the echoes of the screams
in death's silence

we are not alone
what else is there for humanity?
the fragility of life is present
defying the idea that any among us
are beyond death
we find our way to joy
invited to experience
this moment of years
where privilege is overthrown

our unresolved mourning lives
as discomfort and resistance
in the midst of our tenuous existence
we grapple with hope and absence of meaning
turning the traumas of death and empire
into rituals of healing and resistance
creating new presence

conjuring new identities
welcoming new alliances
seeking resurrection as we witness the death of normal

Karen Georgia A. Thompson
14:20
30 October 2020
Olmsted Township, OH

27

Blame the Victim:
When Systemic Injustice
Ceases to be a Culprit

WANDA DEIFELT

Covid was initially treated as the great equalizer, with inspirational statements like 'we are all in this together' and 'we stand strong' in light of adversity. If nothing else, the pandemic caused by the coronavirus has laid bare the social, racial, economic and gender inequalities that plague us also in non-pandemic times. When social distancing was heralded as a necessity to flatten the curve, it also exposed the naked truth that not all human beings are entitled to the same prerogatives: not all workers are able to work remotely, not all have the right to unemployment benefits, not all infected have access to health care, and not everybody has a home to self-quarantine.

Instead of solidarity, Covid aggravated vulnerabilities. A common stratagem in explaining the increasing numbers in contamination and death due to Covid-19 has been to blame the victims. Emphasizing a person's conditions – such as poor health, chronic illness, obesity, homelessness, age etc. – places an onus on the individual without acknowledging the systemic inequalities that lead to these conditions in the first place. Personal vulnerability is labelled as the culprit for massive numbers of casualties due to Covid-19 rather than naming the social and political abandonment that has relegated victims to their historically ascribed place.

The pandemic caused by Covid-19 has also posed questions to theological action and reflection. What can be said and done to address the blatant disregard for human life? How can we move away from religious platitudes and on to a robust, prophetic account of human shortcomings and possibilities of transformative action? Can Jesus' message of love and grace overcome greed and selfishness? Covid has presented itself as an opportunity to assess the human condition and to reflect on our collective human endeavours.

The Human Propensity to Blame Victims

One day, when Jesus was walking along with his disciples, they saw a man who had been blind since his birth. The disciples asked Jesus: 'Rabbi, who sinned, this man or his parents, that he was born blind?' (John 9:1–3a NRSV). The Gospel of John gives us a glimpse of what goes on in the mind of many. The disciples are known for frequently stating the obvious and for playing the role of the average person. But, in doing so, they also offer an opening for a teaching moment, when Jesus – as a brilliant educator (a rabbi) – broadens the horizons of his students and makes them think and act differently.

The question posed by the disciples is the ongoing quest of humanity to find out the reasons for suffering, offering rational explanations for pain and misery. Throughout the centuries, Christian theology has grappled with the subject and offered different readings of theodicy (God's justice, i.e., divine goodness and providence in light of evil).[1] Instead of focusing on the origin of pain, I am interested in the effect suffering has, that is, what human response is to the plight of others. How do we react when confronted with the hurt and anguish of those around us?

I wonder what an equivalent of the biblical scene (John 9:1–3a) would look like today. What would be the reaction of the disciples to the Covid pandemic? Instead of walking along, Jesus and his disciples would need to keep social distance and, through their masks or in a Zoom meeting, they might again ponder: what did the millions of people who contracted Covid-19, and the many who died from it, do to deserve this misfortune? Who sinned?

This analogy might seem frivolous or even outrageous. Who would think of blaming anybody struggling to breathe, wired to a respirator and in complete isolation? It would be simpler to dismiss the comparison altogether but, ironically, this is not far from our reality. The way political authorities and public spokespersons have treated the pandemic – urging businesses to stay open and compelling children to go back to school while seeing the millions of people who get sick and die as 'casualties' – is exactly what 'blaming the victim' looks like.

We learned that Covid is caused by a coronavirus (called SARS-CoV-2) and that older adults and people who have severe underlying medical conditions, such as heart or lung disease or diabetes, are at higher risk of developing more serious complications from Covid-related illnesses. But this assessment is a doubled-edged sword. While it identifies vulnerabilities, it also places a burden on the vulnerable. It seems to excuse the way public health has treated the virus, since the populations at risk are basically bringing illness and death upon themselves, as evidenced in the following CNN interview (Beckett 2020a):

The United States' organized response to the pandemic had been 'historic', Trump's health secretary, Alex Azar, told CNN on 17 May [2020], but America 'unfortunately' has a 'very diverse' population, and black Americans and minorities 'in particular' have 'significant underlying disease'.

Jake Tapper, the CNN anchor interviewing Azar, paused and squinted. Surely, he asked, Azar was not arguing that 'the reason that there were so many dead Americans is because we're unhealthier than the rest of the world?'

Azar doubled down: 'These are demonstrated facts.'

'That doesn't mean it's the fault of the American people that the government failed to take adequate steps in February', Tapper said.

'This is not about fault. It's about simple epidemiology', Azar said, adding in a pious tone: 'One doesn't blame an individual for their health condition. That would be absurd.'

In an attempt to exempt the US government and administration for responsibility for the staggering number of Covid-contaminations and deaths, the Machiavellian solution is exactly that of faulting people for their health condition. This rhetoric blames black Americans for dying from a novel virus because they had diabetes or high blood pressure, in the same way it uses prior conditions to justify access to health care. In other words, it is easier to find fault with somebody's actions than to extend sympathy and show solidarity.

This attitude first dawned on me during the HIV and Aids pandemic. From the outset it was labelled a disease for people at risk because of their risky behaviour. Then, too, the first question asked, when somebody was found to be HIV positive, was: 'How did they get it?' Being gay or drug dependent was an easy way to blame the victim. Some Christians even went as far as to label Aids a punishment from God and a way to call lost sheep back from their devious behaviours.[2] There was no unconditional acceptance. Depending on the answer to the question (how somebody contracted Aids), a person merited empathy or a shrug of shoulders.

Pandemics are times of crisis, and crises reveal both the best and the worst of human behaviour. They lay bare social fissures and our reaction to them. They expose the raw truth that systemic injustices are both overlooked and condoned, by rationalizing the need for sacrifice. In the US, health experts warned that restarting the US economy was premature and risked handing a 'death sentence' to many Americans. One particular political figure argued that older people would rather die than let Covid-19 lockdowns harm the US economy (Beckett 2020b). In doing so, he illustrated the ill-fated logic that some lives are worth saving while others are expendable, and, above all, that profit is above people.

The language of sacrifice is not accidental here. As Henri Hubert and Marcel Mauss explain, a sacrifice is an act of abnegation, depriving the one who is sacrificed: 'Often this abnegation is even imposed upon him as a duty. For sacrifice is not always optional; the gods demand it' (1964, p. 100). The god demanding sacrifice, in this case, is Mammon – the material wealth that debases human life. It shows that some lives do not matter.

When some politicians volunteer people's lives to ensure that businesses and schools stay open, it sends out a message that sacrifice is not only acceptable but required. The inconvenience of wearing a mask has unmasked political bigotry, lockdowns unhinged protests about individual freedoms, and the disproportionate number of brown and black people dying opened wide the wounds of racism. Between saving the economy or people's lives, the increasing number of Covid-cases are described as the lesser of two evils (Noor 2020). This sacrificial ideology reveals the social structures that legitimate power imbalances and justifies the death of those deemed insignificant in society in order to maintain the status quo. Some lives are sacrificed so that others may remain unaffected.

Exposing Vulnerabilities

Indifference is at the heart of the Covid pandemic. Well-being is seen as a merit and access to health (whether admission to medical care and a hospital bed, healthy eating habits, physical exercise, and manageable stress levels) is a reward, not a prerequisite. In countries such as the United States, or Brazil, where I am from, we are taught that collective efforts are subversive (labelled as socialist or communist) and individuals are left to fend for ourselves. The population becomes used to tackling every problem through the lens of 'individual rights' or 'personal privilege', and this disempowers the common good and any collective response to a pandemic – whether it be Covid, HIV-AIDS, racism, sexism or climate change. One of the mechanisms for the perpetuation of social, political, economic, environmental and health challenges is precisely the notion that individual solutions are the answer to collective problems. This makes it possible to blame victims and leave political figures off the hook, or to justify certain actions as a response to their political constituencies as opposed to seeking the common good.

This can be illustrated with the argument employed by the Brazilian president, Jair Bolsonaro, placing Covid-19 in opposition to the welfare of the economy (not of the people), with a willingness to sacrifice human lives to safeguard jobs. In his address to the United Nations in October 2020, Bolsonaro refused any accountability for the chaotic state of the nation. Instead, he attempted to deflect attention by blaming a liberal media that disseminates misinformation and omits reports of the governmental relief to poverty-stricken families because of Covid-19. In his view, Brazil is facing violations against the freedom of religion because Christians are being persecuted (here naming *Cristofobia* or *Christphobia*).[3] A single speech offers a glimpse of the opportunistic politicization of the virus. It conflates criticism of public policies with religious persecution and exempts the greatest beneficiaries of social and economic inequalities from their public responsibility. According to this logic, not only have the ill and dead from Covid-19 brought this upon themselves because of their weakness, but for the physically fit and masculine – which Bolsonaro claims to be – a pandemic-causing virus should be treated as a common cold.

The conflation between health and religion, evidenced in Bolsonaro's speech, is indicative of the rise of prosperity theology in Latin America. Health is a merit and illness is blameworthy. On the one hand, a theology of prosperity capitalizes on the blessings associated with health – it is seen as a divine reward. On the other hand, however, this prosperity is based on a disparity: while the affluent have retreated to private islands and turned lockdown into spa retreats, the majority of the workers – and particularly those in the medical field – had to bear the brunt of illness and death. Here, a theology of prosperity reveals itself as a theology of disparity.

Attitudes such as Bolsonaro's have also fuelled toxic masculinity. The refusal of political leaders to wear a mask and to refrain from public appearance, as did Donald Trump and Boris Johnson, builds on the myth of immunity and pretence of strength even when tested positive for Covid-19.[4] In Bolsonaro's case, there is an ongoing attempt to dismiss the effects of the pandemic and ignore the tragedy of human loss that the virus has unleashed. It has also given him an opportunity to blame victims with the bonus of name-calling. Among his many outrageous public statements, he said that those who become sick are sissies or faggots, that those who take the vaccine may have their DNA altered and become alligators and, ironically, that he wished God's consolation to all those who lost beloved ones due to the virus because he is neither a miracle-worker nor a gravedigger (see Basilio 2020).

It is undeniable that Covid has affected those most vulnerable and opened wide social and racial gaps. In the United States, older black people are more likely to die of the virus than their white counterparts (Evelyn 2020). The struggle with inequities that contribute to premature death rates includes systemic reasons: Poverty and inadequate health care mean higher rates of diabetes, hypertension and heart disease. But it also includes a higher exposure to the virus due to racial inequities in labour (working in essential industries without the necessary protection), population density (higher concentration in urban areas), and overall access to health care (health care is privatized, not easily affordable and can restrict access to treatments).

Covid has revealed a pre-existing pandemic called poverty. As coronavirus ravages the globe, it most brutally impacts poor people and marginalized communities. Hundreds of millions of people are pushed into poverty and unemployment, with an increase in hunger, homelessness and dangerous work. Philip Alston, the outgoing UN special rapporteur on extreme poverty and human rights, summarized this in his final report. Helping the rich getting richer in order to alleviate poverty has entirely upended the social contract and redefined the public good. As a result, it has adopted an 'agenda characterized by deregulation, privatization, lower taxes for corporations and the wealthy, easy movement of money across borders and excessive legal protections for capital' (Alston 2020). The result is that it enriches the wealthy while removing social welfare programmes. Many of the countries that have achieved growth in GDP following the World Bank's poverty measure have also experienced rising hunger, unaffordable health and housing costs, persistent racial wealth gaps, the proliferation of jobs that don't pay a living wage, and ecological devastation.

Covid has also laid bare the gender gap. While job losses due to the virus has affected everybody, it has hurt women in a disproportionate manner. Around the globe, women tend to have the lowest paid jobs in our economy, and women of colour in particular are over-represented in low-wage roles. In a crisis like Covid, the jobs with the lowest pay and less security are often the first ones that employers let go (Connley 2020). In the US, the effect the economic recession created by Covid and the overwhelming effect it has had on female unemployment has been named a 'shecession' – an economic downturn where job and income losses are affecting women more than men (Holpuch 2020).

Worldwide, the pandemic has exposed an increase in incidents of domestic violence:

> Smith, who is also chief executive of a domestic abuse charity [in the UK], said: 'I don't believe coronavirus creates violent men. What we're seeing is a window into the levels of abuse that women live with all the time. Coronavirus may exacerbate triggers, though I might prefer to call them excuses. Lockdown may restrict some women's access to support or escape and it may even curtail measures some men take to keep their own violence under control. (Grierson 2020)

We must recognize that Covid unleashed cultural wars, illustrated by the fact that wearing a simple mask becomes a political statement instead of a responsible and civil act. In 2020, anti-lockdown demonstrations at state capitals showcased a spectrum of views, ranging from anti-government militias and gun-rights advocates to anti-vaccine activists and other conspiracy theorists. Political candidates rallied thousands of people without necessary precautions. Individuals refused to stay at home and attended social gatherings because a lockdown was an inconvenience. When viewing these despicable attitudes, it might seem acceptable to wish the virus on them, or to smile smugly when we receive the news that a political leader who treated the virus as a hoax ends up testing positive for it. This *Schadenfreude* is problematic because it also operates under the logic of assigning misfortune as a result of individual action.

From Structural Sins to Structural Changes

To move away from victim-blaming syndrome requires us to focus on the structural sins that aggravate inequalities and justify individual privilege (Boff and Boff 1987). Naming racism, sexism, classism and ageism as structural sin enables us to address them theologically and not only as sociological markers. When a wrong is accepted, practised and justified within a society it affects everyone, but in particular those who suffer. Jesus' love for the least, the last, and the lost is the motivation for Christian discipleship. In Covid-times, this witness must go beyond much-needed individual acts of kindness to encompass also structural changes in the form of advocacy and greater accessibility to health.

Structural sin is clearly manifested in the unequal access to health care in the US. In the city of Chicago, Covid-19 battered African American communities. In spite of accounting for only 30 per cent of the city's population, black people made up 60 per cent of Covid cases and have the highest mortality rate out of any racial or ethnic group (Oladipo 2020). Wealth and political clout enable some to fly by helicopter to a special wing of a hospital while others cannot even get to an emergency room or a get a simple Covid test. The lack of access to medical treatment is a painful manifestation of structural sin. It demonstrates that there is a social dimension of sin beyond individual wrongdoing. Something has gone horribly wrong in our society when the last words of a patient dying due to Covid-19 are 'who is going to pay for my medical bills?' (Lat 2020)

Social inequality is reflected as much in the spreading of the virus as it is in treating those who are being infected. Access to vaccination is a current example of structural sin, since it will not be readily available to all. The vaccine has brought the health and drug inequalities between richer and poorer countries into sharper focus, with experts warning that less wealthy countries might have to wait until 2024 for a vaccine (Miliband and Gupta 2020). Even before the pandemic, about 20 million infants, often some of the most vulnerable in the world, were missing out on basic vaccines. Roughly half of unvaccinated children live in fragile or conflict-affected countries. Nevertheless, advocating universal access to vaccination can give us the chance to practise justice and solidarity.

A reflection on structural sin leads us to a broader approach and helps to identify alternatives. Take the case of the effect of Covid-19 on children. A simplistic approach might rightfully celebrate that children are much less likely to die from Covid-19. However, the impact of family job loss, disrupted schooling and increased poverty are not so easily avoided. Idrees Kahloon (2020) notes the effect a year without school has on children. The average student would suffer seven months of lost learning, black students would lose ten months, and poor students would forgo the entire year. Poor children do not have reliable internet connection for distance learning, their families cannot pay for supplemental tutoring, and quite often they depend on the school to provide free meals. Covid-19 offers a glimpse of intersectional burdens that require intersectional solutions. Food, housing, parental distress (such as depression and anxiety) and lack of support networks are additional pieces of the well-being puzzle.

Covid has been a wake-up call, getting us ready for even greater challenges than the virus itself. As a result of the pandemic, poverty levels will likely rise to levels unseen. The World Bank predicts that the ranks of the extremely poor – surviving on less than $1.90 a day – will increase by 150 million, and the United Nations asserts that between 240 and 390 million people in 70 countries will be pushed into 'multidimensional poverty' (lacking basic shelter or having children go hungry, for instance; Maslin 2020, p. 74). The social and economic advancements achieved in the last decades will be wiped away in many parts of the world, leaving a dire picture of social inequality and instability ahead.

More than ever the world is in need of solidarity and compassion, offering a counter-narrative to the xenophobic, greedy, and self-entitled behaviours brought out by the pandemic. While Covid exposed the worst in humanity – broadcasting lies and blaming others – it has also shown that human beings can meet the challenge and practise empathy. Instead of propagating bigotry, the coronavirus pandemic also motivated acts of generosity and altruism (Crispin 2020). It is leading us to denounce injustices and call for actions of solidarity.

We are living in a precarious time – a situation of both urgency and vulnerability. As noted by Sturla Stalsett, precarious times 'may constitute a Kairos, a time for decision and action in which faith resources play and important role' (2018, p. 317). Covid has exacerbated inequalities and made visible a global multitude of people who are prevented from life in abundance. They are excluded from life-sustaining and stable labor, accosted by gender and racial discrimination, and stigmatized for physical and mental vulnerabilities. To these, the good news of peace and justice proclaimed by the gospel gains renewed significance.

By Way of Conclusion

The words uttered by the disciples still echo in our own minds: Who sinned? This man or his parents? It is a simpler explanation to blame the victim, find fault with the ill and dead. What did the people who are dying from complications related to Covid-19 do to bring this upon themselves? The majority for being poor, brown and black. For being homeless and sick. For being vulnerable. It is on these bodies that Covid-19 leaves the marks of structural sin. In this, not the blind man nor his parents, all of us share a responsibility. Our shared responsibility refers to the root causes and permanence of poverty and discrimination. It also refers to a shared commitment in finding solutions.

We share a corporate responsibility for sinful actions that originate from social systems because, directly or indirectly, we benefit from them. But then, we must also take our collective responsibility in denouncing these injustices and inequalities, dismantling their luring rhetoric and deadly practices, in order to nurture empathy and solidarity, collective gestures of grace and exercises of justice. Anew, we can reclaim the paradoxical message of the gospel.

Then we can understand Jesus' own words in response to the disciples: 'Neither this man nor his parents sinned,' said Jesus, 'but this happened so that the works of God might be displayed in him. As long as it is day, we must do the works of him who sent me. Night is coming, when no one can work. While I am in the world, I am the light of the world' (John 9:3–5).

Notes

1 Why does God allow evil? This question has had four basic approaches in Christian thought: 1) Some goods are possible only if certain evil exists (this position is defended by Thomas Aquinas, among others); 2) Human freedom is a particularly great good, but free humans may do evil (Alvin Plantinga is a modern example of the 'free will defence'); 3) We become fully human only by being tested against evil (a broad range of theologians adopt this position, ranging from Irenaeus to John Hick); 4) Perhaps there are limits to God's powers (a position adopted by some process theologians). For more details, see Placher, pp. 93–101.

2 An example of this is the reaction of some televangelists, naming hurricane Katrina as God's punishment for gay people in New Orleans (accessed 21.10.20: www.lgbtqnation.com/2016/03/hurricane-katrina-was-gods-way-of-punishing-gay-people/).

3 False claim that Christians are persecuted in Brazil (accessed 22.10.20: www.ihu.unisinos.br/603969-cristofobia-a-farsa-do-opressor-vitimizado?fbclid=IwAR3eqEkdaOe3j7ktqNxRyKMPOKTW57u9O7kM8MzBUAjfTzW19Z_cGY-RoKo).

4 The Covid-19 pandemic has also unleashed a wave of toxic masculinity. As Jackson Katz (2020) points out, politicians such as Mr Trump might not be sophisticated political thinkers, but they understand something fundamental about manhood in a patriarchal culture: the system remains in place because a majority of men fear losing the respect of other men more than they value democracy itself.

References

Philip Alston, 2020, 'Covid-19 has revealed a pre-existing pandemic of poverty that benefits the rich', *The Guardian*, 11 July (accessed 11.7.20: www.theguardian.com/global-development/2020/jul/11/covid-19-has-revealed-a-pre-existing-pandemic-of-poverty-that-benefits-the-rich).

Ana Luiza Basilio, 2020, 'Retrospectiva: as piores declarações de Bolsonaro sobre a pademia', *CartaCapital*, 27 December (accessed 23.3.21: www.cartacapital.com.br/politica/retrospectiva-as-piores-declaracoes-de-bolsonaro-durante-a-pandemia/?fbclid=IwAR1ficu 5i6DPU5G5N3S_ksZM9bkJIL8PUoz3EFiLXIvPyNSHDPk_mHhC_ qE).

Lois Beckett, 2020a, '"All the psychoses of US history": how America is victim-blaming the coronavirus dead', *The Guardian*, 21 May (accessed 9.10.20: www.theguardian.com/world/2020/may/21/all-the-psych oses-of-us-history-how-america-is-victim-blaming-the-coronavirus-dead).

———, 2020b, 'Older people would rather die than let Covid-19 harm US economy – Texas official', *The Guardian*, 24 March (accessed 23.3.21: www.theguardian.com/world/2020/mar/24/older-people-would-rather-die-than-let-covid-19-lockdown-harm-us-economy-texas-official-dan-patrick).

Leonardo Boff and Clodovis Boff, 1987, *Introducing Liberation Theology*, Maryknoll: Orbis.

Courtney Connley, 2020, 'Coronavirus job losses are impacting everyone, but women are taking a harder hit than men', *CNBC*, 14 May (accessed 20.10.20: www.cnbc.com/2020/05/14/coronavirus-job-losses-disproportionately-impact-women.html).

Jessa Crispin, 2020, 'In 2020, Americans helped each other out – because our government wouldn't', *The Guardian*, 28 December (accessed 23.3.21: www.theguardian.com/commentisfree/2020/dec/28/2020-america-coronavirus-covid-politics?utm_term=1b27b2dd3 ae673544e0940ec3635cfc2&utm_campaign=GuardianTodayUS& utm_source=esp&utm_medium).

Kenya Evelyn, 2020, '"The last flag bearers of an era": how coronavirus threatens a generation of black Americans', *The Guardian*, 22 April 39 (accessed 23.3.21: www.theguardian.com/world/2020/apr/21/coronavirus-threatens-generation-of-black-americans).

Jamie Grierson, 2020, 'Domestic abuse killings "more than double" amid Covid-19 lockdown," *The Guardian*, 15 April (accessed 23. 3.21: www.theguardian.com/society/2020/apr/15/domestic-abuse-killings-more-than-double-amid-covid-19-lockdown).

Amanda Holpuch, 2020, 'The "shecession": why economic crisis is affecting women more than men', *The Guardian*, 4 August (accessed

20.1020: www.theguardian.com/business/2020/aug/04/shecession-coronavirus-pandemic-economic-fallout-women).

Henri Hubert and Marcel Mauss, 1964, *Sacrifice: Its Nature and Function*, Chicago: University of Chicago Press.

Idrees Kahloon, 2020, 'The Children of Covid-19', *The Economist*, 13 November, 34.

Jackson Katz, 2020, 'What Donald Trump Understands About American Men', *Ms.*, 22 October (accessed 22.10.20: https://msmagazine.com/2020/10/22/donald-trump-patriarchy-masculinity-men-male-voters/).

David Lat, 'I Didn't Have to Pay a Penny of My $320,000 Covid-19 Hospital Bill. Is That a Good Thing?' *SLATE*, 8 June (accessed 23.10.20: https://slate.com/technology/2020/06/covid-hospital-bill-insurance.html).

Sarah Maslin, 2020, 'The Plague of Poverty', *The Economist*, 13 November, p. 74.

David Miliband and Anuradha Gupta, 2020, 'Covid is a chance to build a world where everyone has access to basic vaccines', *The Guardian*, 18 December (accessed 30.12.20: www.theguardian.com/global-development/2020/dec/18/covid-is-a-chance-to-build-a-world-where-everyone-has-access-to-basic-vaccines).

Poppy Noor, 2020, 'The US politicians volunteering other people's lives to fight Covid-19', *The Guardian*, 22 July (accessed 23.3.21: www.theguardian.com/world/2020/jul/22/us-reopening-politicians-volunteering-peoples-lives-coronavirus).

Gloria Oladipo, 2020, '"It's like they're waiting for us to die": why Covid-19 is battering Black Chicagoans', *The Guardian*, 24 October (accessed 23.10.20: www.theguardian.com/us-news/2020/oct/23/covid-19-battering-black-chicagoans).

William C. Placher (ed.), 2003, *Essentials of Christian Theology*, Louisville: Westminster Press.

Sturla J. Stalsett, 2018, 'Prayers of the Precariat? The Political Role of Religion in Precarious Times', *Estudos Teológicos* 58.2: 313–25.

28

Beyond the Graveyard and the Prison, a New World is Being Born

TINYIKO MALULEKE

The world is in turmoil. Economies are in freefall. We have been brought to our knees – by a microscopic virus. The pandemic has demonstrated the fragility of our world. It has laid bare risks we have ignored for decades: inadequate health systems; gaps in social protection; structural inequalities; environmental degradation; the climate crisis. Entire regions that were making progress on eradicating poverty and narrowing inequality have been set back years, in a matter of months. The virus poses the greatest risk to the most vulnerable: those living in poverty, older people, and people with disabilities and pre-existing conditions. (Guterres 2020)

The Underestimated Game-Changers

In this chapter I argue, among other things, that in light of the ongoing Covid pandemic the time has come for a reconsideration of the game-changing potential of epidemics. I will suggest that epidemics are up there, alongside the wars and the industrial revolutions, as shapers of 'progress', makers of history and disruptors of civilization. They should therefore not be seen as marginal, accidental and rare aberrations. Nor should we see them merely as a recurring but manageable nuisance that temporarily interrupts civilization from time to time. If they were manageable, the world would have long since got rid of them; and if they were merely a nuisance, they would

TINYIKO MALULEKE

not take as many lives as they tend to. Nor would the world expend as many resources – human, infrastructural, scientific and material – on the management and hoped-for eradication of epidemics and infectious diseases.

The initial covid responses tended to be nationalistic and nation-state driven, with country after country issuing targeted travel bans and other inward-looking measures. As a result, some kind of unspoken covid containment competition developed or was sort of declared between countries – with some even creating country ranking tables based on infection and death rates (Howell 2021). Predictably, among the five best performing in such rankings are to be counted some of the smallest countries in the world, with small populations and easy-to-seal borders such as New Zealand, Iceland and Singapore.

For at least half of 2020, many countries were under various levels of lockdown. This was not through a deliberate and globally coordinated effort, but because of the sheer weight of the Covid pandemic touching and interrupting every nation state in the world. With the emergence of 'vaccine nationalism' (*The Guardian* 2021), the same nationalistic approach seems to continue in the search, acquisition and distribution of vaccines.

A Brief South African Excursion

The respected South African historian of epidemics Howard Phillips suggests that such is the global and national impact of epidemics that, among other perspectives, history ought also to be studied from a 'social history of disease angle' (Phillips 2012, p. 11). So far, Phillips (1984 and 1987) has focused on the impact of epidemics on his native land. But his central thesis regarding the historical significance of epidemics is both globally applicable and cross-cutting in its implications. When combined with wars, especially wars of conquest, economic greed and racial prejudice, epidemics have been devastating. Some scholars have attempted a panoramic study of how the Spanish Flu of 1918 changed the world (Spinney 2017) as well as how plagues have shaped the psychology (Taylor 2019),

328

culture and economy of human beings from time immemorial (see also Aberth 2011). Others have employed the literary device of fiction (Garcia Marquez 1988) to understand not only what happens to society when epidemics hit the world (Miller 2011), but what it might be like for individuals to actually live through an epidemic (Camus 1948). There is, understandably, no lack of infectious disease studies, in the broad fields of public health (Cartwright and Biddiss 1980) and history.

In South Africa, the smallpox epidemic of 1713–1893 was a significant factor in the subjugation of the Khoikhoi by the Dutch settlers of the Cape colony.

> Under the weight of these recurring pathogenic blows and the ongoing erosion of their economic, social and political structures by Dutch colonization, Khokhoe autonomy and stability became increasingly compromised and fragile. What tipped the society over the edge was the arrival in 1713 of what in Holland was branded 'the worst of all the harpies; smallpox'. (Phillips 2012, p. 15)

When the plague (1901–07) broke out in South Africa it was not nearly as devastating as the smallpox epidemic had been, but it proved very useful for the powers that be. They used the epidemic to scapegoat the 'natives', to put them under quarantine by force, to forcibly remove them from their residences and to burn down their erstwhile residential areas. Their forced removals under armed guard became the order of the day. Thus were laid the foundations of racial apartheid – half a century before it became official government policy. In Johannesburg,

> the Rand Plague Committee there had all of the 3178 people it had corralled at Coolie Location ... removed under armed guard and sent to a municipal farm at Klipspruit, twelve miles west of Johannesburg, where they were put into tents and huts in two racially separate isolation camps a mile apart, both surrounded by a contingent of troops. ... Back in Johannesburg, the empty Coolie Location was then burnt

to the ground, by, paradoxically, the Johannesburg Fire Department as the soil was deemed too contaminated and the buildings too insanitary to be renovated ... the torching of Coolie Location was 'essentially a theatrical display', calculated to reassure white Johannesburg that no stone would be left unturned to safeguard their health. (Phillips 2012, p. 54)

Similarly, in Port Elizabeth, the white authorities force-vaccinated the local Africans who thereafter had to carry with them documentary proof of vaccination in order to be allowed free movement. All over South Africa, Africans, Indians and Coloureds were forced out of towns and settled further away from towns. As a result, new 'townships' which were hastily and artificially put together, emerged: Klipspruit (later renamed Pimville and Soweto) in Johannesburg, New Brighton in Port Elizabeth, Ginsberg in East London and Ndabeni (later called Cape Flats) in Cape Town. Before the formation of the Union of South Africa (1910) and even before the introduction of the infamous Land Act of 1913 which effectively made it illegal for Africans to own land (Plaatje 1916), white authorities across the land were using the plague as an excuse to evacuate Africans from the cities and to control their movement. In this way, pandemics and epidemics have always had many 'uses'.

Epidemics in the Era of Sustainable Development Goals

The UN ideal of disease eradication has proved elusive. Yet, in light of the enormous cost in lives, livelihoods and economies, one wonders whether epidemic disease readiness ought not to have been included among the millennial development goals. Strictly speaking, the world can never be 100 per cent ready for the next epidemic, but we can be more prepared than we were in 2019 when the novel coronavirus started wreaking havoc, first in China before spreading to the rest of the world.

In our days, the global agenda for development is enshrined in the Sustainable Development Goals (SDGs), which were

BEYOND THE GRAVEYARD AND THE PRISON

adopted on 25 September 2015 by all members of the United Nations. While the third (out of 17) SDG speaks of the quest for healthy lives, the reduction of child mortality and the need to keep raising life expectancy, it steers away from mentioning epidemics. Similarly, while poverty and hunger are singled out for eradication among the SDGs, the language relating to health spells out no ambition for the elimination of disease as such.

The global ambition for the eradication of disease and epidemics is located in the mandate of the WHO. The landmark objective of WHO, as captured in the first article of its constitution, is 'the attainment by all peoples of the highest possible level of health' (WHO 2005). Consequently, among the 17 strategies (called 'functions') in the WHO constitution, which are designed to assist the organization in the pursuit of its overarching objective, is the goal 'to stimulate and advance work to eradicate epidemic, endemic and other diseases' (WHO 2005).

Part of the familiar narrative of linear human progress, that is, the idea that the quality of human life has been incrementally improving through the centuries, was the idea of a world free of diseases. And yet, no sooner are announcements made about the eradication of one epidemic or the other than a new one or a new strain emerges.

Despite this, the disease eradication narrative continues in new guises. Declaring the age of epidemics and pandemics already passé, Yuval Noah Harari claimed that it is likely that 'major epidemics will continue to endanger humankind in the future only if humankind itself creates them, in the service of some ruthless ideology. The era when humankind stood helpless before natural epidemics is probably over' (2015, p.14). But Harari also makes the argument that unlike the twentieth, the twenty-first century is not the century of the masses – mass education, mass medicine, mass military personnel; instead the twenty-first century is the age of elite superhumans and algorithms. As a result of this, the mission of medicine has changed so that whereas 'twentieth century medicine aimed to heal the sick, twenty first century medicine is increasingly aimed to upgrade the healthy' (Harari 2015, p. 348).

So while Harari noted that there are reports of new bacteria that are becoming resistant to well-known antibiotics, so that even the gains of the much hailed era of antibiotics which started in the 1950s are not so secure he insists that this is a temporary situation, which will soon be resolved once and for all, by science. And yet, some viral diseases of epidemic proportions have become part and parcel of our lives. For this reason, we take all the necessary and available precautions such as Malaria pill regimes and our Yellow Fever vaccination certificates when visiting affected geographical regions. One hundred and three years since the Influenza pandemic, the yearly flu vaccine notwithstanding, most people have come to expect and to accept their annual bout of flu at the applicable time of the year. In fact, the common flu still takes a considerable number of lives per annum, globally. But if the utterances of the likes of Yuval Noah Harari are anything to go by, it should not be long before either the whole world or at least the world of the elite superhuman minority, lives in a post-disease, post-epidemic world.

A World Made in the Image of a Pandemic

Who can use the term 'gone viral' now without shuddering a little? Who can look at anything any more – a door handle, a cardboard carton, a bag of vegetables – without imagining it swarming with those unseeable, undead, unliving blobs dotted with suction pads waiting to fasten themselves on to our lungs? Who can think of kissing a stranger, jumping on to a bus or sending their child to school without feeling real fear? (Roy 2020)

Indeed, since Covid-19 broke out, we fret and fuss about a microscopic virus that lies waiting to ambush us at the shop floor, in smoke-filled night clubs where we used to let our hair down to dance the night away, on the supermarket trolley handles, inside lifts, in the air that we breathe, in the breath of strangers and in the spittle of friend and foe. But over, above and in between the paranoia that has come to characterize the people of the world at these times, something else is happening. Covid-19 is *changing* not only the way we live but also the way we feel, hear, see, think and imagine. It is also *revealing* the usually disguised, hidden, tolerated and the taken-for-granted injustices and prejudices of human society. In the words of António Guterres, the pandemic can be

> likened to an X-ray, revealing fractures in the fragile skeleton of the societies we have built. It is exposing fallacies and falsehoods everywhere: The lie that free markets can deliver healthcare for all; the fiction that unpaid care work is not work; the delusion that we live in a post-racist world; the myth that we are all in the same boat; because while we are all floating on the same sea, it's clear that some of us are in super-yachts while others are clinging to the floating debris. (Guterres 2020)

In comparing the impact of Covid-19 to an X-ray image of human society in all its nakedness, so that it is reduced to its bare bones, UN Secretary General António Guterres cuts to the heart of the matter. Occurring as it does at the height of liberal economics, stubborn racism (Maluleke 2020c), and the hype (Harari 2015) about the fourth industrial revolution (Schwab 2017), Covid-19 has put a few question marks on some of the dominant narratives of our time. Essayist and novelist Arundhati Roy described the palpable impact of Covid-19 in India in the following terms:

> The lockdown worked like a chemical experiment that suddenly illuminated hidden things. As shops, restaurants, factories and the construction industry shut down, as the wealthy and the middle classes enclosed themselves in gated colonies, our towns and megacities began to extrude their working-class citizens – their migrant workers – like so much unwanted accrual. Many driven out by their employers and landlords, millions of impoverished, hungry, thirsty people, young and old, men, women, children, sick people, blind

people, disabled people, with nowhere else to go, with no public transport in sight, began a long march home to their villages ... Some died on the way. (Roy 2020)

Two powerful metaphors from two influential thinkers of our time regarding the impact of Covid-19: an X-ray image of the world (Guterres) and a chemical experiment that suddenly illuminated hidden things (Roy). The image of the exodus of the poor after the expulsion from the cities of India is itself a powerful metaphor of the fate of the poorest of the poor all over the world. Not that the virus distinguishes between classes, genders and 'races'; only that it worsens the already worse situation of the poor and marginalized. In South Africa the government exhorted everyone to frequently wash hands with soap, to sanitize, to keep social distances and to observe the lockdown by staying at home. The problem is that in many areas of the country, the poor have no homes to speak of, and no water and no sanitation. In the midst of the pandemic, images of poor South African villagers were seen rummaging for water in the sand where there once was a brook. Many had never seen a 'sanitizer' before, nor could they afford buying it or buying masks. In the dozens of shanty towns and squatter camps across the country, social distancing has proven to be very hard. The cultural impact (Maluleke 2020e) of Covid19 prevention and containment protocols (Maluleke 2020a) have been particularly hard on the poor who, by reason of their poverty, live in close proximity, making social distancing near impossible. The rich, with their large gardens and their sprawling ranches, practise social distancing permanently, in a way. The social distancing protocols that applied to many national lockdowns suited the rich very well. But they were no less vulnerable to the virus. Until there is enough and effective vaccination for 'herd immunity' to be attained globally, no one is invulnerable.

During the second wave of the Covid pandemic the body count has risen to alarming levels in many countries. What has become clear is that the whole world now lives in a set of 'new normal' conditions. So much of what had been taken for granted is up in the air. The quality of relations between human beings, between humanity and other creatures, and between human beings and the environment, is being thoroughly tested.

Theological Reflection

Encouraged by the historical insights from Howard Phillips and the metaphors used by the likes of Guterres and Roy, and in order to bring some theological reflection as I move towards the conclusion of the chapter, I would like to lean on two other metaphors – the world as a graveyard and the world as a prison. The first builds on the story of the man with an unclean spirit, as told in Mark 5 (see Maluleke 2009). The second is borrowed from the latest novel by British-Nigerian writer, Ben Okri (see Maluleke 2019).

My Fellow Graveyard Men and Women

More than a decade ago, I published a Bible study based on Mark 5 (Maluleke 2009) on which I wish to build this section. Mark 5 begins with the story of a man possessed with an impure spirit – who lived in a graveyard by the sea, who we have renamed 'graveyard man'. To prepare his reader for the looming drama, Mark (5:3–5) first provides, in the fewest of words, an abridged CV, and a vivid character mapping of the 'graveyard man'.

> This man lived in the tombs, and no one could bind him any-more, not even with a chain. For he had often been chained hand and foot, but he tore the chains apart and broke the irons on his feet. No one was strong enough to subdue him. Night and day among the tombs and in the hills he would cry out and cut himself with stones.

As Jesus arrives in the Gerasene region by boat, the troubled man emerged out of a graveyard and pounced. He went run-ning towards Jesus and knelt before him. Before the man could utter a single word, Jesus issued a command for the impure spirit to get lost. But the unclean spirit inside the graveyard man responded and proceeded to conduct a curious conversation with Jesus. 'What do you want with me, Jesus, Son of the most High God? In God's name don't torture me!' pleaded the unclean spirit.

I would suggest that the seemingly disagreeable utterances of the graveyard man must be taken with a pinch of salt. Had he really wished to be left alone, he could have just stayed in the 'comfort' of the graveyard that was his habitat! After all, is the graveyard not the perfect place to learn how not to feel? Part of the problem with the crucial and life-saving Covid-19 preventive protocols – particularly social distancing and the prohibition of bodily contact – is that they also have the effect of moving us gently into a social graveyard and an emotional void, unless we develop alternative rituals of affection, solidarity and empathy.

Five things are striking about the graveyard man: 1) He lived in the graveyard. 2) He had extraordinary physical power – he tore chains apart and broke iron shackles. 3) He felt no pain – night and day he cut himself with stones. 4) He had no name – he called himself 'legion' in reference to the many spirits which lived inside him, meaning he had an identity crisis. 5) He walked about naked or near-naked – so that after he was healed he was found sitting, dressed and in his right mind.

I would like to suggest that these striking characteristics of the graveyard man and his context resemble our own, in the covid 'new normal' world. We too have long insulated ourselves against pain, the pain of others and our own pain. We too have an identity crisis: are we creatures or are we creators (so called *homo deus*)? Since sickness and death will, allegedly, be defeated soon, are we mortal or are we immortal, are we human or are we just a bundle of inferior algorithms? We are naked too. The coronavirus has stripped human society naked – to the bone.

Some of our countries have the awesome but useless power of the graveyard man. Consider the military might of our most powerful nations – take, China, the USA and North Korea, for example. If the coronavirus was a military foe, they would have long sent targeted drones to go and blow it to smith-ereens. Similarly, if the coronavirus was merely a malevolent algorithm, you can bet Huawei, Google, Samsung, Microsoft and Apple would have built a leaner, meaner and faster algorithm to eat up the coronavirus algorithm. For more than a year, all the military power, all the artificial intelligence and all the technological power and – for a moment – all the monetary power in the world had no answer to the wrecking ball called Coronavirus. As US Secretary General António Guterres has observed, despite the awesome power – military, technological and monetary – concentrated in the hands of a few superpower countries, 'we have been brought to our knees – by a micro-scopic virus'. Not only has the power of the most powerful among the nations of the world been of no use in the face of the coronavirus, but this power is also a lot more efficient in destruction, including self-destruction.

Living in the Graveyard

At my most depressed moments during the South African lock-down, which started in the middle of March 2020 and spilled over into 2021, the reality of living in a graveyard hit me. What with so many friends, colleagues and relatives dying at a time when we were not able to perform the normal mourn-ing rituals. Reports that the South African government had secured large tracts of land in order to dig more than a million graves (Reimann 2020; see also eNCA 2020), in preparation of anticipated Covid deaths, seemed to confirm that my country was becoming a giant graveyard. And so, in my mind, I went looking for the graveyard man of Mark 5.

When I walked to the local pharmacy, which suddenly inflated the prices of masks, sanitizers and face shields, I was on the lookout for the graveyard man. Even when I undertook the many 'pilgrimages' from my study to the kitchen, from the front door to the small garden at the back, from the bedroom to the bathroom and all the way back, I would stop to look into the mirror. Until one day, lo and behold, the man in the mirror was none other than the graveyard man himself, staring back at me. There was a smirk on his face.

In those days, it felt as if my entire country was the graveyard. Actually, I imagined a massive and unstoppable graveyard, extending across the land, taking over the country, acre by acre, one square kilometre at a time like the shadow of a moving cloud. In this daytime nightmare of mine, our four major cities – Johannesburg, Pretoria, Cape Town and Durban – were the pulsating epicentres from which the giant graveyard was daily expanding seismically, across the land, in a thousand concentric circles of death.

To live in the graveyard is to live in the midst of and side by side with death. To live in the graveyard is to make peace with death, not merely to accept its inevitable approach, but to actually meet death halfway.

My Fellow Inmates

Early in 2019 the celebrated Nigerian writer Ben Okri published his latest novel – a dystopian text of considerable force and terror (see Maluleke 2019). The setting for the drama of the novel is a 'new' world. That whole world is a giant prison in which people live according to the most brutal and the most ridiculous rules and regulations. In that world, thinking is a criminal offence. Books have been banned for so long, people have forgotten how to write and make books. Sleeping, including sleepwalking, is mandatory. So is wailing in the night. The people are losing their minds – those with residual minds to lose, that is. In this world the rich minority live in luxury and the poor majority scramble for bread in the streets.

But here's the thing: despite all these absurd and dictatorial regulations, the secretive and elusive Hierarchy that rules the world does not allow anyone to see, depict or name the world as a prison. Instead, the Hierarchy does everything in its power – and it had all power – to suppress all (propagation of) past and present knowledge of the world as a prison. The concern is that the more often the populace hears about and sees portrayals of the world as prison, they may realize that indeed, they do live in a prison. In that case, they may wake up from their legislated slumber, and rise in protest, demanding freedom.

Such is the determination of the ruling Hierarchy to suppress all intimations about the world as a prison they incentivize all alternative depictions and portrayals of the world – especially the world as a garden rather than a prison. Writers, singers and artists are encouraged and paid to produce alternative artistic depictions of the world – away from the image of the world as a prison. But behind all these benevolent gestures in support of depictions of a different world, there was an iron fist. This came in the form of a vicious police squad, much like the notorious Special Anti-Robbery Squad of Nigeria (Malumfashi 2020). But unlike the infamous SARS of Nigeria, the riot squad under the control of the Hierarchy in Ben Okri's book, specialized not so much in shooting down the people. It was a cannibalistic riot squad that specialized in devouring the protesters live – literally. On standby to assist the cannibalistic police squad in their murderous orgy were the lords and the baronesses of the land.

There is no way that Ben Okri could have predicted the Covid fiasco! But his novel is instructive in several ways. In its own way, Covid has turned the world into a giant prison. Millions of people were imprisoned in their countries, cities, villages and homes. Under cover of the lockdown enforcement, the police and the army of some African countries, like South Africa and Nigeria went rogue.

Allegations soon emerged that in the course of enforcing the lockdown, the South African army and the police (Maluleke 2020d) had caused the deaths of at least eleven South Africans, the first one killed within days of the very first proclamation of lockdown. The killing of George Floyd in the USA resonated strongly with the experiences of black South Africans. By the end of July 2020, the South African media revealed a massive corruption scandal (Maluleke 2020b) in terms of which money set aside as relief for the impact of Covid for the poor was nefariously redirected into the bank accounts of political leaders and/or their collaborators.

Like the cannibalistic police squad in Ben Okri's novel, corrupt politicians, trigger happy members of the police and the army, joined forces and started cannibalizing their fellow South Africans through corruption and violence.

Concluding Thoughts: Moving Beyond the Gateway

If my country, South Africa, was a microcosm of the world, then we must say that each pandemic we have experienced over the past three centuries has had a most disruptive effect on social arrangements and economic prospects alike. However, we must quickly add that, in the experience of South Africa, such disruptions have seldom been of such a nature as to fundamentally alter the living conditions of the poor for the better, let alone shift power from the powerful to the powerless.

Over the past 300 years South Africa has suffered four major health epidemics – smallpox, bubonic plague, poliomyelitis, HIV/AIDS – and now we are in the midst of a fifth pandemic, Covid. While testing of the social fabrics, public health and economic structures of the country, none of the four health pandemics and epidemics that we have endured as a nation over the past 300 years has resulted in the improvement of the lot of the poor.

To conclude, let me return to Arundhati Roy and borrow yet another metaphor for the 'new normal world' we should seek – the metaphor of a portal and a gateway.

> Nothing could be worse than a return to normality. Historically, pandemics have forced humans to break with the past and imagine their world anew. This one is no different. It is a *portal*, a gateway between one world and the next. We can choose to walk through it, dragging the carcasses of our prejudice and hatred, our avarice, our data banks and dead ideas, our dead rivers and smoky skies behind us. Or we can walk through lightly, with little luggage, ready to imagine another world. And ready to fight for it. (Roy 2020)

Come then, my friends, let us not settle at this portal/gateway that is located between the old and the new. To settle here would be to make the desert traversed by the children of Israel, and not the promised land, the destination of our exodus. In the words of Nelson Mandela, 'we dare not linger' here, for our 'long walk is not yet ended' (Mandela 1994, p. 751). What then is the world we seek? We seek a world in which good news will be proclaimed to the poor and not to the super-human elite minority that rules the world; a world in which all manner of prisoners will be set free; a world in which the oppressed will be freed from the yoke of poverty – a world in which the blind will recover their sight (see Luke 4.18).

References

John Aberth, 2011, *Plagues in World History*, Lanham, MD: Rowman & Littlefield.

Albert Camus, 1948, *The Plague*, London: Hamish Hamilton.

F. F. Cartwright and M. D. Biddiss, 1980, *Disease and History: The Influence of Disease in Shaping the Great Events of History*, London: The History Press.

eNCA, 2020, 'Gauteng prepares one million graves' (accessed 20.1.21: www.youtube.com/watch?v=StU19013cCE).

Gabrel Garcia Marquez, 1988, *Love in the Time of Cholera*, New York: Alfred A. Knopf.

The Guardian, 2021, 'The Guardian view on vaccine nationalism: think again', *The Guardian*, 29 January (accessed 23.3.21: www. theguardian.com/commentisfree/2021/jan/28/the-guardian-view-on-vaccine-nationalism-think-again).

António Guterres, 2020, Nelson Mandela Annual Lecture, 18 July (accessed 20.1.21: www.nelsonmandela.org/news/entry/annual-lec ture-2020-secretary-general-guterress-full-speech).

Yuval Noah Harari, 2015, *Homo Deus: A Brief History of Tomorrow*, London: Harvill Secker.

Beth Howell, 2021, 'The Countries Who've Handled Coronavirus the Best – and Worst', *Movehub*, 29 January (accessed 26.1.21: www. movehub.com/blog/best-and-worst-covid-responses/).

Tinyiko Maluleke, 2009, 'The Graveyard man, the "escaped convict" and the girl child: A mission of awakening, an awakening of mission', *The International Review of Mission* 91: 550–7.

————, 2019, Ben Okri's 'The Freedom Artist: A Vision of Hell on Earth', *Sunday Times* [South Africa], 26 March (accessed 28.1.21: www.timeslive.co.za/sunday-times/books/news/2019-03-26-ben-okris-the-freedom-artist-a-vision-of-hell-on-earth/).

————, 2020a, 'Love in the time of Covid-19', *News24*, 22 March (accessed 15.1.21: www.news24.com/news24/Columnists/Guest Column/opinion-love-in-the-time-of-covid-19-20200322-2).

————, 2020b, 'My fellow South Africans, the Season of Weeping is Upon Us', *Sunday Independent*, 2 August (accessed 23.3.21: www.iol.co.za/sundayindependent/dispatch/my-fellow-south-africans-the-season-of-weeping-is-upon-us-96799fd3-49c2-449e-8b9a-10a3b94cbc3c).

————, 2020c, 'Racism *En Route*: An African Perspective', *Ecumenical Review* 72.1: 19–36.

————, 2020d, 'Say their names, President Ramaphosa', *Sunday Independent*, 7 June (accessed 28.1.21: www.iol.co.za/sundayindependent/dispatch/say-their-names-president-ramaphosa-49082804).

————, 2020e, 'The social and cultural implications of Covid-19' *News24*, 13 March (accessed 12.1.21: www.news24.com/news24/columnists/guestcolumn/opinion-the-social-and-cultural-implications-of-covid-19-20200313).

Sada Malumfashi, 2020, 'Nigeria's SARS: A brief history of the Special Anti-Robbery Squad', *AlJazeera*, 22 October (accessed 29.1.21: www.aljazeera.com/features/2020/10/22/sars-a-brief-history-of-a-rogue-unit).

Nelson Mandela, 1994, *Long Walk to Freedom*, London: Little Brown Company.

Andrew Miller, 2011, *Pure*, London: Hodder & Stoughton.

Howard Phillips, 1984, 'Black October: The Impact of the Spanish Influenza Epidemic of 1918 on South Africa', PhD thesis, University of Cape Town.

————, 1987, 'Why did it Happen? Religious and lay Explanations of the Spanish Flu Epidemic of 1918 in South Africa', *Kronos* 12: 72–92.

————, 2012, *Plague, Pox and Pandemics*, Johannesburg: Jacana.

Sol Plaatje, 1916, *Native Life in South Africa*, London: King and Son.

Nicholas Reimann, 2020 'South Africa Readies 1.5 Million Graves for Coronavirus Mass Burials', *Forbes*, 8 July (accessed 23.3.21: www.forbes.com/sites/nicholasreimann/2020/07/08/south-africa-readies-15-million-graves-for-coronavirus-mass-burials/?sh=7971ab4a45a6).

Arundhati Roy, 2020, 'The Pandemic is a Portal', *Financial Times*, 3 April (accessed 25.1.21: www.ft.com/content/10d8f5e8-74eb-11ea-95fe-fcd274e920ca).

Klaus Schwab, 2017, *The Fourth Industrial Revolution*, London: Penguin.

Laura Spinney, 2017, *Pale Rider. The Spanish Flu of 1918 and How it Changed the World*, London: Jonathan Cape.

Steven Taylor, 2019, *The Psychology of Pandemics: Preparing for the Next Global Outbreak of Infectious Disease*, Newcastle-upon-Tyne: Cambridge Scholars Publishing.

WHO, 2005, *Constitution of the World Health Organization* (accessed 15.1.21: https://apps.who.int/gb/bd/PDF/bd47/EN/constitution-en.pdf?ua=1).

29

new hope

blaming the victim
blaming groups of people who are unwell

empire eating the lives of God's people away
disease eating away the lives of God's people
what do we have to say?

God's silence is perplexing
disease a mystery to be solved

requiring intervention of a Divine presence
where is God in the messy complications
of pandemic and propensity to oppress the poor

safety of the world
arbitrarily rendered

refusals to support the lives and living of the masses
masks pushed aside as illegitimate response
competing needs interrupting the possibilities of healing

where to lay blame as the suffering increases
where to find answers that include a Gospel of grace
where to place our hope in these days of death
when graveyards are walking into the seas

a scramble for good live
destruction of abundant living

fighting for attention from a God co-opted
culpable religion supporting gods of capitalism
oppression named god

the earth begs rest in an unforeseeable future
where empire is toppled

nation states grapple with diminishing resources
the opportunity to heal the masses
confused with the search for a cure

the competition identifies a price at the end
a capitalist need for empire

to once again
exhibit dominance
increasing revenue at the price of those in need
teetering on the margins

competition that defies the cry for help
blaming those who are impacted by disease

here in a race for millions at stake
the world becomes smaller
the earth a prison turning graveyard

borders become prison walls
boundaries mark graveyards

visions needed to tear down these walls
vision for healing a world diagnosed
with apathy and uncompassionate malaise

learning escapes the masses
signs point to need for change

systems begging for alternatives
where is the liberation of God promised
in a state of life where capitalism still prospers

the silence drives us into deeper necro-politics
the margins refuse the return to normal

emerging are new ways for living this life
emerging is the fight for new life

life for the poor
life for the unwell
life for the un-dead

new life
new hope
new freedoms

Karen Georgia A. Thompson
13:25
1 November 2020
Olmsted Township, OH

Acknowledgements

The author and publisher are grateful for permission to use extracts from the following sources.

Ulrike Bail, 'In search for a place for those who died alone', in *Statt einer ankunft*, St. Ingbert, Germany: Conte Verlag (2021). Used by permission.

Danny Dziuk, 'The bright blue sky of these days', at www.youtube.com/watch? v=2-5sc8OzfDk. Used by permission.

John Robert Lee, 'Collage', in *Pierrot*, Loods: Peepal Tree Press Ltd (2020). Used by permission.

Index of Names and Subjects